Semi-Field Methods for the Environmental Risk Assessment of Pesticides in Soil

Other Titles from the Society of Environmental Toxicology and Chemistry (SETAC)

Ecotoxicology of Amphibians and Reptiles
Sparling, Linder, Bishop, Krest, editors
2010

Ecological Assessment of Selenium in the Aquatic Environment
Chapman, Adams, Brooks, Delos, Luoma, Maher, Ohlendorf, Presser, Shaw, editors
2010

Application of Uncertainty Analysis to Ecological Risks of Pesticides
Warren-Hicks and Hart, editors
2010

Risk Assessment Tools Software and User's Guide
Mayer, Ellersieck, Asfaw
2009

Derivation and Use of Environmental Quality and Human Health Standards for Chemical Substances in Water and Soil
Crane, Matthiessen, Maycock, Merrington, Whitehouse, editors
2009

Linking Aquatic Exposure and Effects: Risk Assessment of Pesticides
Brock, Alix, Brown, Capri, Gottesbüren, Heimbach, Lythgo, Schulz, Streloke, editors
2009

Aquatic Macrophyte Risk Assessment for Pesticides
Maltby, Arnold, Arts, Davies, Heimbach, Pickl, Poulsen
2009

Ecological Models for Regulatory Risk Assessments of Pesticides: Developing a Strategy for the Future
Thorbek, Forbes, Heimbach, Hommen, Thulke, Van den Brink, Wogram, Grimm, editors
2009

For information about SETAC publications, including SETAC's international journals, Environmental Toxicology and Chemistry and Integrated Environmental Assessment and Management, contact the SETAC office nearest you:

SETAC
1010 North 12th Avenue
Pensacola, FL 32501-3367 USA
T 850 469 1500 F 850 469 9778
E setac@setac.org

SETAC Office
Avenue de la Toison d'Or 67
B-1060 Brussells, Belguim
T 32 2 772 72 81 F 32 2 770 53 86
E setac@setaceu.org

www.setac.org
Environmental Quality Through Science®

Semi-Field Methods for the Environmental Risk Assessment of Pesticides in Soil

Andreas Schaeffer
Paul J. van den Brink
Fred Heimbach, Simon P. Hoy
Frank M.W. de Jong, Jörg Römbke
Martina Roß-Nickoll
José P. Sousa

SETAC Workshop PERAS
Coimbra, Portugal

Coordinating Editor of SETAC Books
Joseph W. Gorsuch
Copper Development Association, Inc.
New York, NY, USA

CRC Press
Taylor & Francis Group
Boca Raton London New York

CRC Press is an imprint of the
Taylor & Francis Group, an informa business

CRC Press
Taylor & Francis Group
6000 Broken Sound Parkway NW, Suite 300
Boca Raton, FL 33487-2742

First issued in paperback 2017

© 2011 by Taylor and Francis Group, LLC
CRC Press is an imprint of Taylor & Francis Group, an Informa business

No claim to original U.S. Government works

ISBN 13: 978-1-138-11796-9 (pbk)
ISBN 13: 978-1-4398-2858-8 (hbk)

Visit the Taylor & Francis Web site at
http://www.taylorandfrancis.com

and the CRC Press Web site at
http://www.crcpress.com

SETAC Publications

Books published by the Society of Environmental Toxicology and Chemistry (SETAC) provide in-depth reviews and critical appraisals on scientific subjects relevant to understanding the impacts of chemicals and technology on the environment. The books explore topics reviewed and recommended by the Publications Advisory Council and approved by the SETAC North America, Latin America, or Asia/Pacific Board of Directors; the SETAC Europe Council; or the SETAC World Council for their importance, timeliness, and contribution to multidisciplinary approaches to solving environmental problems. The diversity and breadth of subjects covered in the series reflect the wide range of disciplines encompassed by environmental toxicology, environmental chemistry, hazard and risk assessment, and life-cycle assessment. SETAC books attempt to present the reader with authoritative coverage of the literature, as well as paradigms, methodologies, and controversies; research needs; and new developments specific to the featured topics. The books are generally peer reviewed for SETAC by acknowledged experts.

SETAC publications, which include Technical Issue Papers (TIPs), workshop summaries, newsletter (*SETAC Globe*), and journals (*Environmental Toxicology and Chemistry* and *Integrated Environmental Assessment and Management*), are useful to environmental scientists in research, research management, chemical manufacturing and regulation, risk assessment, and education, as well as to students considering or preparing for careers in these areas. The publications provide information for keeping abreast of recent developments in familiar subject areas and for rapid introduction to principles and approaches in new subject areas.

SETAC recognizes and thanks the past coordinating editors of SETAC books:

A.S. Green, International Zinc Association
Durham, North Carolina, USA

C.G. Ingersoll, Columbia Environmental Research Center
US Geological Survey, Columbia, Missouri, USA

T.W. La Point, Institute of Applied Sciences
University of North Texas, Denton, Texas, USA

B.T. Walton, US Environmental Protection Agency
Research Triangle Park, North Carolina, USA

C.H. Ward, Department of Environmental Sciences and Engineering
Rice University, Houston, Texas, USA

Contents

List of Figures

List of Tables

Acknowledgments

This book presents the proceedings of the workshop "Semi-field Methods for the Environmental Risk Assessment of Pesticides in Soil" (PERAS), convened by the Society of Environmental Toxicology and Chemistry (SETAC) Europe in Coimbra, Portugal, in October 2007. Fifty-two scientists involved in this workshop represented 17 countries from Europe, the United States, Brazil, and Canada and offered expertise in ecology, ecotoxicology, environmental regulation, and risk assessment. Where relevant, further developments and literature on the workshop topics since 2007 have been included (see Chapter 1).

The workshop was made possible by the generous support of the German Federal Environment Agency (UBA), the Dutch Ministry of Housing, Spatial Planning and the Environment (VROM), the IMAR-Coimbra Interdisciplinary Centre, BASF, Bayer CropScience, and Syngenta (Appendix 3).

The workshop also was supported by the excellent management of the local organizing committee led by José Paulo Sousa (assisted by Sónia Chelinho and Xavier Domene) and the general organizing committee consisting of Martina Roß-Nickoll, Fred Heimbach, Simon Hoy, Frank de Jong, Jörg Römbke, José Paulo Sousa, and Andreas Schäffer (chair). Thanks also to Paul van den Brink, who summarized the workshop discussions as a workshop rapporteur.

We further gratefully acknowledge many valuable and helpful comments of several colleagues during preparation of the document, especially Björn Scholz-Starke and Bernhard Theißen (both RWTH Aachen University), responsible for the chapter classification of existing semi-field approaches, and Thorsten Leicher (Bayer CropScience), responsible for the draft method description on terrestrial model ecosystem (TME) tests. Also, we appreciate the supporting comments of Anne Alix as external reviewer.

About the Editors

Andreas Schäffer works as professor for environmental biology and chemodynamics at RWTH Aachen University, Germany. In 1984, he earned his PhD in chemistry at Münster University, Germany, and studied biochemistry as a postdoctoral fellow at Harvard University, Boston, and Zürich University, Switzerland. He worked in the agrochemical industry from 1989 to 1997 to study the environmental fate of pesticides in soil and water. Research of his university group focuses on the environmental risk assessment of pollutants in soil, plants, and water. He is especially interested in the ecochemistry (degradation, binding, and transport) and ecotoxicology (terrestrial and aquatic) of xenobiotics, and he develops strategies for bioremediation of polluted sites using plants. He serves as member of environmental risk assessment working groups (German Ministry of User Safety and Food Safety [BVL], European Food Safety Authority [EFSA]), and as board member of environmental organizations (SETAC German Language Branch and GDCh, Ecochemistry/Ecotoxicology division).

Fred Heimbach works as a consultant scientist at RIFCon GmbH in Leichlingen, Germany. He obtained his MSc degree and PhD in conducted research on marine insects at the Institute of Zoology, physiological ecology, at the University of Cologne. From 1979 until 2007, he worked at Bayer CropScience in Monheim, Germany, on the side effects of pesticides on non-target organisms. In addition to his work, he has given lectures on ecotoxicology at the University of Cologne. Heimbach has researched the development of single-species toxicity tests for both terrestrial and aquatic organisms and has worked with microcosms and mesocosms in the development of multispecies tests for these organisms. As an active member of European and international working groups, he participated in the development of suitable test methods and risk assessment of pesticides and other chemicals for their potential side effects on non-target organisms. For several years, he has served on the SETAC Europe Council and the SETAC World Council, and he has been an active member of the organizing committees of several European workshops on specific aspects of the ecotoxicology of pesticides.

Frank de Jong works as a senior risk assessor at the National Institute of Public Health and the Environment (RIVM) in Bilthoven, the Netherlands. In 1987 he obtained his MSc in biology at Leiden University; in 2001 he obtained his PhD in environmental sciences at Leiden University. Between 1987 and 2001, he worked at the Institute of Environmental Science of Leiden University, where he did a large number of research projects aimed at agricultural-environmental problems, especially pesticides, and a PhD study aimed at terrestrial field trials assessing the side effects of pesticides. From 2001 to present, de Jong has worked at the RIVM, where his main field of interest is the evaluation of higher-tier studies submitted in the framework of pesticide registration, and the development of methods for summarizing and evaluating terrestrial and aquatic (semi)field studies. He is the secretary of the Dutch Platform for the Assessment of Higher Tier Studies. Apart from this, he is involved in the derivation of environmental quality standards and advisory work for the NL government in the field of risk assessment of pesticides.

Simon Hoy is a senior scientist working for the Chemicals Regulation Directorate (CRD) in York, United Kingdom. The CRD is the United Kingdom's national regulatory authority for plant protection products (PPPs), biocides, and other chemicals; its PPPs section was formerly known as PSD. Hoy joined PSD in 1990, following his interest and education in entomology, agricultural zoology, and conservation biology (Wolverhampton, Leeds, and Exeter Universities, 1982 to 1990). Currently Hoy's main role is as a regulatory ecotoxicologist and environmental risk assessor for new and existing chemicals and products, although he also advises on regulatory and environmental policy and research for chemicals at a national and international level. He has worked closely with other European member states and various European Union authorities conducting environmental risk assessments for PPPs and developing related guidance and guidelines. He has a particular interest and experience in invertebrate ecotoxicology and soil risk assessment and has previously been actively involved in European and Mediterranean Plant Protection Organization (EPPO) and SETAC workgroups and workshops in this field.

Jörg Römbke is managing director of ECT Oekotoxikologie GmbH, a private contract research laboratory located in Floersheim, Germany. He obtained his diploma in zoology and doctorate in biology from the University of Frankfurt in 1983 and 1988, respectively. Between 1981 and 1993 he worked in the Department of Applied Biology at Battelle Institute in Frankfurt, focusing his activities on monitoring anthropogenic stress on soil ecosystems, mainly forests. In 1993 he co-founded ECT GmbH. Since then Römbke has acted as senior scientist in this company, being active in areas such as development and validation of terrestrial test methods, performance of standardized tests for the registration or notification of chemicals, ecological risk assessment of the side effects of these compounds, and basic research on soil ecology in European and tropical ecosystems. In addition to his work he is actively participating in test method standardization, for example, as a member of ISO Technical Committee 190.

Martina Roß-Nickoll works as senior scientist and group leader for terrestrial ecology and ecotoxicology at the Institute of Environmental Research at RWTH Aachen University. In 1999, she earned her PhD in biology at RWTH Aachen University. From 1999 until 2000 she worked in the student advice center of RWTH Aachen University. Her research on the ecology of biocenoses aims to assess terrestrial communities depending on site-specific properties. Data from monitoring of plants and animals (arthropods in particular) at selected sampling sites are combined into classification schemes by employing multivariate statistical methods. Based on these data, an improved assessment of ecological consequences of external impacts on communities and biodiversity will become possible. Since 2003 she has been a board member of the research institute for Environmental Analysis and Assessment gaiac, Aachen, Germany. She is the immediate past president of the SETAC German Language Branch.

José Paulo Sousa is a professor at the Department of Zoology, University of Coimbra (UC), and a senior researcher at IMAR–Coimbra Interdisciplinary Centre. Since 1993 he has been developing research on soil ecology and soil ecotoxicology. On soil ecology, his activities are focused on the development and validation of ecological and biodiversity indicators (using soil invertebrates) to assess changes in soil quality derived from changes in land use in agricultural and forest systems. Regarding soil ecotoxicology, he has been working mainly in the Mediterranean region and in the tropics (primarily in South America) focusing on the development and adaptation of bioassays with soil organisms to assess effects of chemical substances and wastes on soil quality, and to integrate ecological risk assessment (ERA) schemes of contaminated soils. He is also involved in the development of new approaches to assess effects of pesticides on soil-water interfaces. He was responsible for the elaboration of test guidelines with soil invertebrates edited by the International Organization for Standardization (ISO) and the Organization for Economic Cooperation and Development (OECD). Currently he is a member of the Ecotoxicology Working Group of the EFSA PPR Panel. Since 2005, he has been acting as an external consultant for EMBRAPA Agrobiologia (Brazil) for soil ecology and soil ecotoxicology issues.

Paul van den Brink, workshop rapporteur, is a professor of chemical stress ecology and works at the research institute Alterra and the Aquatic Ecology and Water Quality Management Group of Wageningen University, both belonging to the Wageningen University and Research Center. He is the immediate past president of SETAC Europe and an editor of the SETAC journal *Environmental Toxicology and Chemistry*.

Workshop Participants

Ahtiainen, Jukka
OECD/Finnish Environment Institute/
    SYKE, Finland

Amorim, Monica
University of Aveiro, Portugal

Candolfi, Marco P.
BASF AG, Crop Protection Division,
    Germany

Cerejeira, Maria José
Instituto Superior de Agronomia,
    Portugal

Coulson, Mike
Syngenta, United Kingdom

de Jong, Frank
National Institute for Public Health and
    the Environment, Netherlands

Dinter, Axel
DuPont de Nemours (Deutschland)
    GmbH, Germany

Egsmose, Mark
EFSA, Italy

Fontes, Tânia
DGADR, Portugal

Förster, Bernhard
ECT Oekotoxikologie GmbH, Germany

Frampton, Geoff
University of Southampton, United
    Kingdom

Frische, Tobias
UBA, Germany

Garcia, Marcos
Embrapa Amazônia Ocidental, Brazil

García-Gómez, Concepción
INIA, Department of Environment,
    Laboratory for Ecotoxicology, Spain

Gottesbüren, Bernhard
BASF AG, Crop Protection Division,
    Germany

Hammers-Wirtz, Monika
gaiac, Germany

Hardy, Ian
Battelle UK Ltd., United Kingdom

Heimbach, Fred
Bayer CropScience AG, Germany

Hommen, Udo
Fraunhofer Institute for Molecular
    Biology and Applied Ecology,
    Germany

Hoy, Simon
Chemicals Regulation Directorate,
    United Kingdom

Klimczak, Katarzyna
Instytutu Ochrony S'rodowiska w
    Warszawie, Poland

Knäbe, Silvio
eurofins-GAB GmbH, Germany

Kölzer, Uschi
Bayer CropScience AG, Germany

Krieg, Wolfgang
BASF AG, Crop Protection Division,
Germany

Kula, Christine
BVL, Germany

Kuperman, Roman
Edgewood Chemical Biological Center,
United States

Laszkowski, Rychard
Jagiellonian University, Poland

Lawlor, Peter
Pesticide Control Service, Ireland

Leicher, Thorsten
Bayer CropScience AG, Germany

Lührs, Ulf
IBACON GmbH, Germany

Miles, Mark
Dow Agrosciences, United Kingdom

Möbes-Hansen, Britta
AGES, Institut für Pflanzen-
schutzmittelbewertung und
-zulassung, Austria

Nienstedt, Karin
European Food Safety Authority
(EFSA), Italy

Nikolakis, Alexander
Bayer CropScience AG, Germany

Ratte, Hans Toni
RWTH Aachen, Institute for
Environmental Research, Germany

Redolfi, Elena
ICPS, Italy

Römbke, Jörg
ECT Oekotoxikologie GmbH, and
BiK-F Research Centre, Germany

Roß-Nickoll, Martina
RWTH Aachen, Institute for
Environmental Research, Germany

Ruf, Andrea
University of Bremen, Germany

Schaeffer, Andreas
RWTH Aachen, Institute for
Environmental Research, Germany

Scholz-Starke, Björn
RWTH Aachen, Institute for
Environmental Research, Germany

Scott-Fordsmand, Janeck
National Environmental Research
Institute, Denmark

Scroggins, Rick
Environment Canada, Canada

Sousa, José Paulo
University of Coimbra, Department of
Zoology, Portugal

Stupp, Hans-Peter
Bayer CropScience AG, Germany

van Beelen, Patrick
National Institute for Public Health and
the Environment, Netherlands

van den Brink, Paul
Alterra and Wageningen University,
Netherlands

van der Geest, Bert
FURS, Slovenia

van Eekelen, Renske
CTGB, Netherlands

van Vliet, Peter
CTGB, Netherlands

Verstraete, Ann
Ministère de l'Agriculture, Belgium

Vergnet, Christine
AFSSA, France

Photo: Wolfgang Krieg

Workshop participants at work.

Executive Summary

Current Europe Union regulations (European Council Directive 91/414/EEC 1991, to be replaced by the upcoming Regulation 1107/2009) for the authorization of plant protection products (PPPs) require an assessment of the effects of pesticides on soil organisms (such as earthworms) and soil functions (such as microbial respiration and breakdown of soil organic matter) through a tiered approach, from laboratory to semi-field tests, the latter being applied as higher-tier methods when lower-tier methods fail to adequately address the risk. To discuss experiences with various semi-field methods, including terrestrial model ecosystems (TMEs), and their potential role as higher-tier tools, an expert workshop was organized under the auspices of SETAC Europe. The workshop PERAS ("Semi-field Methods for the Environmental Risk Assessment of Pesticides in Soil") was held in Coimbra, Portugal, in 2007. The output from this workshop, condensed into these proceedings, provides guidance as to criteria that need to be assessed when considering conducting higher-tier semi-field studies, recognizing that each assessment and choice of method is approached on a case-by-case basis.

The aims of the PERAS Workshop were as follows:

- To highlight the current state of knowledge regarding semi-field methods and to identify the most appropriate methods to assess the impact of chemicals on soil community structure and function (Chapters 2 and 3)
- To give a particular focus on higher-tier laboratory and semi-field methods, including TMEs, which may be employed between first-tier laboratory tests and full-scale field studies (Chapter 4)
- To discuss technical aspects of the TME method in order to agree, as far as possible, on a standardized test method (Chapter 5)
- To identify key gaps in knowledge and areas for further research and development in testing the effects of PPPs in soil (Chapter 6)

Because of the considerable experience of several researchers in the use of TMEs, a section of the workshop focused on this method. While TMEs represent possibly the most frequently used semi-field method for assessment of effects of pesticides on soil-inhabiting species, it was emphasized that other semi-field approaches may also be employed, depending upon the nature of the issue being assessed.

The workshop debated 4 technical areas (key points are summarized next), which were considered as critical guidance for the design and conduct of semi-field tests, including some points specifically relevant for TME studies.

1) Environmental fate and exposure considerations
   - Readily degradable pesticides (DT90 < 100 days) should be applied to soil semi-field systems according to good agricultural practice (GAP).

- The application of persistent pesticides should also take into account the formation of an accumulation plateau concentration in the soil.
- In order to enable extrapolation of the test results to other environmental conditions, on both a spatial and a temporal scale, different soils, soil moistures, seasons, etc., should be tested (ideally during method development).
- Recommendations for ranges of soils for semi-field studies should be established.
- Ideally analytical confirmation of the added pesticide concentrations should be achieved in the same soil strata that will be sampled for the ecotoxicological effects assessment.

2) Effect considerations
- TMEs might appropriately mirror field situations, provided that potentially sensitive soil organism groups are present at sufficient abundance within the cores and that they are exposed to the test substance.
- Soil micro- and mesofauna (e.g., nematodes, microarthropods, enchytraeids) and microorganisms are the main groups targeted by this particular system.
- If TMEs are used, they need to be left to stabilize for a period before treatment. Grassland soils are preferred because they provide richer and more stable biocenoses than other land uses.
- Existing experience with TMEs supports sizes between 10 and 50 cm in diameter, and about 40 cm in depth.
- A range of functional and structural endpoints can be measured.
- To evaluate the suitability of semi-field test systems with respect to the sensitivity of the soil biota, toxic (positive) controls should be included in the study design.
- The magnitude and duration of effects can be measured in replicated semi-field systems.

3) Sampling
- The sampling design of a semi-field study is driven largely by the characteristics of the test substance, in particular its fate in soil. One must differentiate between nonpersistent and persistent chemicals. With nonpersistent chemicals, sampling efforts should focus on the beginning of the study and on potential recovery periods following application.
- In the case of persistent chemicals, a longer assessment timeframe is required.
- Considering the experience gained with earthworm field studies, samples should be taken before the start of the study and at least 4 times after application of the test substance up to a period of 1 year.
- The size of the test system will determine what can be studied; for example, the decomposition of organic matter can be studied in litter bags in the field while the use of pieces of cellulose paper would be more appropriate in TMEs.
- If TMEs are used, it is not possible to make a final recommendation about the pros and cons of taking several subsamples from the same TME as opposed to destructively sampling the whole TME. In a semi-

field plot study, appropriate distances between subsamples can be built into the design.

4) Statistical considerations

- Variability within TME studies can be high due to the patchy distribution of soil organisms; however, they are probably no more variable than results obtained from field studies.
- Variability due to environmental conditions or heterogeneity within soil cores taken from the field can be minimized by the careful choice of the test site and prescreening of the patchiness of organism distribution and soil properties.
- A prospective power analysis based on existing data on the variables within a semi-field system such as a TME may be of help in order to determine how many replicates are needed to achieve a desired minimal detectable difference (MDD). For example, 80% power to detect 50% deviation of a treated plot or TME from a control plot or TME could be a suitable threshold.
- Statistical analysis depends on the test design, which may range from a dose-response experiment to a limit test at a single concentration. Appropriate numbers of replicate treatments will be governed by the statistical power of the analysis undertaken.
- Both uni- and multivariate statistical methods could be applied to improve understanding, interpreting, and determining the validity of the study. Univariate methods describe the change in abundance of populations, whereas when applying multivariate methods, for example, principal response curves (PRCs), the alteration of community structure may be derived.
- The whole data set of a semi-field test should be interpreted by experts with a sound knowledge of the biology and ecology of the soil biota.

Various ecological and performance criteria need to be met when selecting a particular semi-field approach for a higher-tier effects assessment study.

ECOLOGICAL CRITERIA

Relevance: The system should include important species (e.g., ecosystem engineers, keystone species, sensitive species), or in case of multispecies tests systems, the species composition should represent the community of the habitat of concern.

Endpoints: Total number, covering structure and function.

Flexibility: Suitable for different exposure scenarios, different soil types, and different crops.

Sensitivity: Sensitive to chemicals but robust toward other factors. The system should react in a relevant dose range. The system should contain sensitive species.

PERFORMANCE CRITERIA

Practicability: Good ratio between resources (costs, time, staff) and results.
Reproducibility and repeatability: Statistical robustness, that is, low variability of chosen endpoints.
Experience: Amount of studies performed, including field comparisons.
Standardization: Guideline or guidance paper available.

KEY RECOMMENDATIONS

Several different approaches for higher-tier testing are available, including semi-field methods such as TMEs (with the most experience being gained so far), which may offer a range of potential tools for higher-tier environmental risk assessment of pesticides in soil.

- The selection of the most appropriate higher-tier method (laboratory, semi-field, or field) should be guided by the research or risk assessment requirements and the regulatory question that needs to be addressed.
- When a semi-field study is designed, account should be taken of the ecology of the key species under investigation and the fate and behavior characteristics of the test substance.
- For the assessment of the acceptability of effects found in semi-field studies, a clear definition of soil protection goals is needed. From these protection goals, the level of protection can be deduced, and from this, the need for and suitability of a semi-field test to show the magnitude and duration of certain effects can be determined. For acceptance of such an approach at the European level, which may then be translated into regulatory guidance, the PERAS workshop participants proposed that a further workshop be organized to determine the appropriate protection goals for agricultural soil.

Extended Summary

BACKGROUND

Beside many other factors, pesticides may lead to spatial and temporal changes in soil biological communities in the agricultural landscape.

The scope of this guidance from the SETAC workshop "Semi-field Methods for the Environmental Risk Assessment of Pesticides in Soil" (PERAS), Coimbra, Portugal, 2007 (see Appendix 1 for the program), is to identify and describe suitable semi-field test methods that are able to detect potential effects of pesticides on soil communities in the tiered approach of pesticide risk assessment within the European Union. Fifty-five experts from academia, industry, and authorities, for example, the European Food Safety Authority (EFSA), Organization for Economic Cooperation and Development (OECD), and national pesticide registration agencies, were invited from Europe, Brazil, and the United States to discuss the state of the art with a focus on semi-field methods, including terrestrial model ecosystems (TMEs).

Based on a review of existing semi-field approaches, the PERAS proceedings focus on tests consisting of intact soil cores with natural communities (TMEs), as defined later. The potential for the use of TMEs in pesticide risk assessment was mentioned in the European and Mediterranean Plant Protection Organization (EPPO) risk assessment scheme for soil organisms and functions in 2003 and also in the "Guidance Document on Terrestrial Ecotoxicology" under Council Directive 91/414/EEC (EC 2002). While TMEs were mentioned as a potential higher-tier refinement step, it was not clear precisely how such methods would fit into a tiered risk assessment scheme. The use of semi-field methods, including TMEs, may gain importance with the upcoming Regulation 1107/2009 replacing Directive 91/414/EEC, where the focus of soil risk assessment is shifting away from effects on soil function and more toward "structural" and biodiversity effects, in particular soil community structure.

ECOLOGICAL CONSIDERATIONS

Soil provides a medium for an astounding variety of organisms that use the soil as a habitat and a source of energy and contribute to the formation of soil by influencing the soil's physical and chemical properties and the nature of vegetation that grows on it. The 5 interacting soil-forming factors are the parent material, climate, relief, biota, and time. Natural and anthropogenic factors will lead to spatial and temporal changes in soil biological communities.

In the context of the PERAS proceedings, the information provided covers mainly temperate regions of the northern hemisphere, with a focus on the agricultural landscape, comprising crops, meadows, orchards, grassy field margins, hedges, floodplains, etc.

When discussing soil organism communities, it is important to distinguish between the structure and the function of the soil biocoenosis or community at the ecosystem level:

- Structure is the composition of the soil biocoenosis (biodiversity), described at the species level (i.e., abundance, biomass, diversity, and dominance).
- Function refers to biologically determined processes in the soil ecosystems, based on the interaction of its different components (i.e., nutrient cycling, community respiration, organic matter breakdown, stabilization, and structuring of soil aggregates).

Despite a high organism variability in time and space as well as differences in sampling methods used, rough estimates of the soil biodiversity indicate several thousands of invertebrate species apart from the largely unknown microbial and protozoan diversity: By far the most dominant groups of soil organisms, in terms of numbers and biomass, are bacteria and fungi. Besides these organisms, soil ecosystems generally contain a large variety of animals, such as protozoa, nematodes, microarthropods such as mites, collembola, or oligochaetes such as enchytraeids and earthworms. In addition, a high number of macrofauna species, mainly arthropods, are living in the uppermost soil layers, the soil surface, and the litter layer, which are organized at different trophic levels and groups in the soil food web. The use of the landscape by humans has brought about profound changes aboveground and belowground, and even provoked global changes. In this context, knowledge of soil structure and functional processes becomes increasingly relevant, but the assessment of any interaction between structure and function in the field often suffers from the lack of a reference site and a good understanding of the role of individual soil organisms.

LEGISLATIVE AND REGULATORY BACKGROUND TO THE ASSESSMENT OF RISKS FROM PLANT PROTECTION PRODUCTS IN SOIL

The regulation of plant protection products (PPPs) in the European Union (EU) until 2009 was undertaken according to European Commission Directive 91/414/EEC (European Council Directive 91/414/EEC 1991) to be replaced by Regulation 1107/2009 (EC2009) and its various annexes and amending directives. The ecotoxicology data requirements generally cover both acute and chronic effects and follow a tiered testing and risk assessment framework; that is, they start with relatively simple acute laboratory tests and move on to more complex chronic laboratory, semi-field, and field tests. While semi-field methods might well prove useful in assessing effects on, for example, some surface active arthropods and plants, the focus here will be on those non-target organisms that predominantly inhabit the soil profile: earthworms, microarthropods, soil microorganisms, and other soil non-target macroorganisms (which are used in the directive to predict effects on the breakdown of soil organic matter).

In the EC directive, only the risk to earthworms is determined by directly considering effects on the organisms themselves, that is, by testing and assessing "structural" effects on a single species in the laboratory (including lethal and sublethal effects), or on population and community structure and diversity under field conditions. Effects on soil microorganisms are determined largely through tests only on the processes of microbially mediated carbon and nitrogen mineralization in the laboratory, although the possibility of field testing remains. While other soil macroorganisms might be tested, research has suggested that no clear relationship can currently be established between structural effects on individual species and likely impacts on soil functions such as soil organic matter breakdown, so this link is intuitive rather than proven.

Reasons for the increased recent focus on discussing semi-field methods are

1) the lack of a suitable internationally agreed, validated, and standardized (semi-field) test method for studying the structural aspects of soil communities;
2) the upcoming Regulation 1107/2009 that will replace Directive 91/414/ EEC, raising the need for higher-tier methods to address structural aspects; and
3) a clear need for the definition of protection goals.

For different aspects of pesticide registration a clearer definition is needed of what the regulatory testing procedure is intended to protect, under what circumstances, and what level of impact or effects might be considered acceptable. Besides these aspects, it should be clearly defined what magnitude and extent of effects any new test method is able to detect, how recovery can be studied, and whether the method is able to be adapted to different protection goals.

Many participants at PERAS considered that the current regulatory position, while not ideal in some respects, has evolved into a reasonably well-understood testing and risk assessment strategy for PPPs. There was some understandable resistance to modify this position, while there was no significant movement or change in the perceived protection goals for soil. PERAS itself was not seen to be an appropriate forum to determine this direction, which would be developed by other working groups and forums. Nevertheless, the question of whether the protection goals for soil should be predominantly structural or functional, or both, was still intensively discussed at PERAS. While the functional endpoints might be sufficient to protect the key processes in soil that enable it to retain its utility as an agricultural resource, they provide little information on the impact on individual taxa, populations, or communities of organisms. Clear distinctions were apparent among delegates at PERAS, some of whom considered that within the field environment, which is already severely impacted by operations such as plowing, the functional integrity of the soil was the ultimate protection goal. Others felt that the in-field soil community was still part of the whole agricultural landscape and required protection of its structural diversity, particularly from longer-term effects and persistent compounds. Whether structural or functional, there was, however, agreement at PERAS that the main protection goal (in-field at least) should be for recovery or recolonization from adverse effects to occur within 1 year (or cropping season) following initial pesticide application. Any semi-field method would, therefore, need to run sufficiently long enough to predict this recovery.

Further guidance on the question of whether both structural and functional effects are relevant protection goals is established by the upcoming Regulation 1107/2009 that will repeal Directive 91/414/EEC (along with its various annexes). The relevant EC guidance documents, which describe in more detail the risk assessment procedures to follow, are also due to be revised. At present, views expressed during the revision of the ecotoxicological data requirement Annexes II and III to the directive indicate that the risk assessment for soil should be approached in a more integrated way, and include tests on structural endpoints of in-soil communities, rather than relying on the soil microorganism tests and other functional endpoints. If the regulatory requirements develop toward a greater focus on soil biodiversity and community structure for a wider range of soil fauna (and possibly still functions) than currently considered, there is a clear scope for the increased practical use of semi-field methods. Such methods will, however, need to remain flexible and adaptable to meet whatever regulatory question arises from lower-tier studies.

Except for trigger values indicating persistence in soil, there is no broadly accepted evaluation procedure at the European level for the evaluation of persistence in soil, and member states use different approaches at the national level. The evaluation procedure proposed in the Netherlands for persistent plant protection products in soil (in-crop area) was provided. The proposal itself is not a product of the PERAS workshop and was recognized by the participants as 1 national example of how protection goals might be set, how terrestrial higher-tier methods might be used, and risks determined.

In order to render assessment possible, certain protection goals are suggested in the Dutch proposal. Because the PERAS workshop also underlined the need for soil protection goals, the Dutch proposal was offered as a starting point for discussion about how such protection goals might appear and be put into practice.

According to the Dutch proposal, the functional redundancy principle (FRP) aims at the protection of "life support functions" of the in-crop soil to allow the growth of the crop and protection of key(stone) species (earthworms) of agricultural soils. This protection goal is, in effect, already assessed at the European level according to existing PPP requirements and it is not discussed further here. The community recovery principle (CRP) aims at protection of life support functions of the soil to allow crop rotation and sustainable agriculture, with overall protection of the structure and functioning of soil community characteristics for agroecosystems. The ecological threshold principle (ETP) aims at protection of life support functions of the soil to allow changes in land use, with overall protection of the structure and functioning of soil community characteristics for nature reserves. The procedure provides trigger values for the half-life for dissipation (DT50) from soil. Separate decision schemes were proposed for each protection goal. In these schemes both the predicted environmental concentrations (PECs) and the ecotoxicological endpoints can be determined using tiered approaches.

A tiered assessment is proposed for effects in the risk assessment. As a first-tier approach it is proposed to base the permissible concentration on the long-term toxicity exposure ratio (TER) for a basic set of standard soil organisms. This base set is different for compounds with different modes of action; for example, for insecticides a soil test with arthropods should be included. As a second tier, the species sensitivity distribution (SSD) approach is proposed calculating the hazardous concentration for 5% of the species (HC5). At the highest tier of the refined effects assessment, the performance of semi-field tests is proposed.

Also in the Dutch proposal, "effect classes" can be used to facilitate the interpretation of concentration-response relationships for relevant measurement endpoints in terrestrial semi-field experiments, that is, from Class I (no treatment-related effects) up to Class V (clear long-term effects; full recovery not within 1 year of the last application of the PPP in the test system). In the Dutch proposal these effect classes were mainly based on experience in the aquatic environment, because there is a lack of data for the terrestrial environment.

One of the conclusions of the Dutch proposal is that protocols for higher-tier studies are lacking, and the development of semi-field methods is recommended. In a worked example using carbendazim, the available literature concerning TME studies is used for higher-tier risk assessment.

OVERVIEW AND EVALUATION OF SOIL
SEMI-FIELD (HIGHER-TIER) METHODS

Semi-field tests are defined as controlled, reproducible systems that attempt to simulate the processes and interactions of components in a portion of the terrestrial environment, either in the laboratory (small scale) or in the field, or somewhere in between. They are designed in a way that the advantages of laboratory tests (e.g., standardization, controlled conditions) are combined with the advantages of field studies (natural variability, complex interactions), while at the same time avoiding their disadvantages, like simplicity or high amount of man power, respectively.

The following typology for the various semi-field tests is proposed:

- Group A methods comprise artificially assembled units with added organisms under controlled environmental conditions (A1) or under field conditions (A2).
- Group B tests are TMEs using intact soil cores with natural communities under controlled environmental conditions (B1) or under field conditions (B2).
- Field enclosures on undisturbed soil belong to group C, in which immigration of species is prevented by barriers; tests can be performed relying on natural communities (C1) or on added organisms (C2), respectively.

In addition, "grey zones" exist between the laboratory and the semi-field level, and between the semi-field level and the field level. A literature review focused on ecotoxicological methodological papers: In total, approximately 150 papers were identified, including a high number of "grey" reports which—in part—still need to be assessed. Out of those, 51 papers on semi-field methods were evaluated in detail. From these, 34 papers focused on assembled soil systems (group A) (70%), 10 papers on TMEs (group B) (20%), and 7 papers on field enclosures (group C) (10%).

Examples of A1 tests, so-called "gnotobiotic tests," are those in vegetation chambers, terrestrial microcosm chambers, microecosystems (MESs), integrated soil microcosms (ISMs), the MS•3 test, or the soil multispecies test system (SMS). Intact soil cores with natural communities (TMEs) under laboratory conditions belong to group B1, developed more than 20 years ago and still the only standardized terrestrial semi-field method. However, the same TME approach under field conditions uses larger soil columns (up to 47 cm in diameter, 40 cm in height, 100 kg of grassland soil). It has been shown in these tests that population dynamics of most dominant species follow natural fluctuations over the years. There are only a few examples known for field enclosures with natural communities (C1); more common are tests with added organisms (C2).

In order to evaluate the suitability of the presented semi-field tests to be used as a higher-tier system in the context of the risk assessment of pesticides, 2 groups of criteria are described, covering aspects of ecology and performance. Since it is difficult to quantify the degree of fulfillment of these criteria, it was decided to use expert knowledge for this evaluation process. There is no "best use" method for all purposes of higher-tier testing of chemicals, and it has to be decided case by case which method is favorable. For instance, if an impact on 1 important predator species is of interest, a field enclosure may be an option. On the other hand, if structural endpoints are the main focus and indirect effects along the food web are expected, a TME study may be preferable. Always a balance has to be struck between the greater repeatability, precision, and relative simplicity of laboratory studies, and the greater ecological realism but increasing complexity and costs of semi-field and field studies.

In summary, based on a thorough and intense literature review, a number of suitable semi-field methods are available for the environmental risk assessment of pesticides in soil. The pros and cons of the different methods were discussed in detail during the workshop. In the context of proposed regulatory requirements, the TME approach is considered an appropriate semi-field method to resolve questions resulting from lower-tier assessment studies.

TECHNICAL GUIDANCE

The discussions held in 4 "technical" working groups at the PERAS workshop are summarized next. Because by far the most experience available related to TMEs, the discussion during the workshop mainly focused on this method. However, this does not preclude that the experience gained with TMEs, or the recommendations made during the workshop, are not suitable for other semi-field methods.

FATE AND EXPOSURE CONSIDERATIONS

Readily degradable pesticides (DT90 < 100 days) should be applied to soil semi-field systems according to good agricultural practice (GAP); however, the application of persistent pesticides should also take into account the formation of an accumulation plateau concentration in the soil. Soils in semi-field tests should always be covered with plants (at least following application), and irrigation of the soil should be adapted to regional circumstances and according to the water demands of the cover crop and soil fauna. It was debated during the workshop whether persistent pesticides should be applied to agricultural (i.e., arable) soils only, while nonpersistent pesticides might be applied to either grassland or arable soils. However, this should be considered on a case-by-case basis. Studies addressing a risk assessment for in-crop concerns may need a different approach than studies relating to off-crop concerns because of differences in soil cover. In order to enable extrapolation of the test results to other environmental conditions, on both a spatial and a temporal scale, different soils, soil moistures, seasons, etc., should be tested (ideally during method development). Recommendations for ranges of soils for semi-field studies should be established.

To evaluate the suitability of semi-field test systems with respect to the sensitivity of the soil biota, toxic (positive) controls should be applied to separate test systems. Ideally, analytical confirmation of the added pesticide concentrations should be achieved in the same soil strata that will be sampled for the ecotoxicological effects assessment.

EFFECT CONSIDERATIONS

There was a consensus that TMEs might appropriately mirror field situations, provided that potentially sensitive soil organism groups are present at sufficient abundance within the cores and that they are exposed to the test substance. It was recognized that soil micro- and mesofauna (e.g., nematodes, microarthropods, enchytraeids) and microorganisms are the main groups targeted by this particular system. A pre-application incubation time is needed (flexible, but varying from 1 to several weeks) allowing communities to adapt and stabilize within the TME after the disturbance caused by the extraction of soil cores. TMEs may potentially be applicable to crop areas, grasslands, or even forests. However, existing experience shows a strong preference for grasslands due to their higher diversity and stability of soil organism communities. The selected size of a TME is rather a compromise between technical effort and ecological relevance. Size is also a function of the sampling strategy adopted, that is, sub-sampling or destructive cores. Existing experience with TMEs support sizes between 10 and 50 cm in diameter, and about 40 cm in depth. Soil moisture content was identified as the most important soil property, meaning that it should be measured and compensated regularly. More research is needed in order to clarify the pros and cons of outdoor vs. indoor TMEs. Micro- and mesofauna community structure is the often most relevant effect parameter, that is, species number, dominance structure, abundance, and trophic structure. Usually, the duration of a TME study is 16 weeks, and current experience indicates that they have a limited life span of up to 1 year. Functional endpoints as well as microbial activity (e.g., soil enzymes), respiration, and functional diversity can also be assessed in TME studies. Moreover, integrative functional parameters like litter decomposition or the feeding rate via bait-lamina can also be measured. Finally, the classification of the magnitude and duration of effects in TME studies could be performed in a way similar to that of schemes already developed for aquatic mesocosms.

SAMPLING

The sampling design is mainly driven by the characteristics of the test substance, in particular its fate in soil. In particular, one has to differentiate between nonpersistent and persistent chemicals. In the former case, sampling efforts should focus at the beginning of the study and on potential recovery periods following application. In the case of persistent chemicals, a longer assessment timeframe is required. In any case, information from lower tiers should be used extensively in order to help identify relevant sampling endpoints and time points. According to the experience gained with earthworm field studies, samples should be taken before the start of the study and at least 4 times after application of the test substance up to a period of 1 year. Technically speaking, sampling does not differ considerably in TMEs from those methods recommended for field studies. For example, samples for mesofauna are taken with a soil corer, but the surface of a TME is usually not big enough to expose litter bags, meaning that decomposition of organic matter could be tested with pieces of cellulose paper. Currently, it is not possible to make a final decision about the pros and cons of taking several subsamples from the same TME vs. sampling the whole TME destructively at each sampling date. Because it can be expected that TME studies will usually be performed in soils that can be utilized for agriculture (including meadows), difficulties with very sandy or clayey soils will be the exception.

STATISTICAL CONSIDERATIONS

Variability within TME studies can be high due to the patchy distribution of soil organisms; however, they are probably no more variable than results obtained from field studies. Variability due to environmental conditions/heterogeneity within soil cores taken from the field can be minimized by the careful choice of the test site and prescreening of the patchiness of organism distribution and soil properties.

Before a TME experiment is started, a prospective power analysis based on already existing data on the variabilities may be of help in order to determine how many replicates are needed to achieve a desired minimal detectable difference (MDD). It was discussed that, for instance, 80% power to detect 50% deviation of a treated TME from a control TME could be a suitable threshold.

Statistical analysis depends on the test design: From a dose-response experiment an ELx or ECx can be derived for the test substance; if a no-observed-effect level or concentration (NOEL or NOEC) has to be derived, the statistical power will strongly depend on the number of replicates. When performing a limit test with finite resources, more replicates are possible at the highest treatment rate to increase the power at that pesticide concentration. Both uni- and multivariate statistical methods could be applied to improve understanding, interpreting, and determining the validity of the study. The type of methods used will lead to various interpretation possibilities: Using univariate methods, the change in abundance of populations can be described; applying multivariate methods, for example, principal response curves (PRCs), the alteration of community structure may be derived. It was agreed in the workshop that the whole data set of a semi-field test needs interpretation by experts with sound knowledge of the biology and ecology of the soil biota.

RESEARCH NEEDS

Research needs for semi-field test methods are summarized from those discussed during the PERAS workshop:

1) Protection goals and risk assessment scheme: A clear definition of soil protection goals is needed, from which the level of protection can be deduced, and from this the ability of a test to show the magnitude and duration of certain effects. Ideally, a workshop should be organized, to which risk managers and decision makers are particularly invited, to determine the appropriate protection goals for agricultural soil. (Note: Since the PERAS workshop, this is likely to be taken forward during revision of the EC terrestrial and persistence guidance documents.)

2) Basic ecological research: Literature research as well as experimental work are necessary to describe the ecology, sensitivity against PPPs, and recovery timeframes for mesofauna for both in-crop and off-crop communities. The complex dependencies between soil communities and landscape heterogeneity (soil properties, vegetation pattern, and land use) should be classified using a reference system.

3) Uncertainties in extrapolation: Data from semi-field tests need to be extrapolated to other environmental conditions, such as different soil types (texture, pH, cation exchange and water holding capacities, organic matter content, nutrient status), as well as temperature, water (irrigation vs. natural climatic conditions), and light regimes. Little is known about the variability of data from semi-field tests in comparison to data from field tests. Therefore, comparative experimental studies and models are considered necessary to determine first the variability of TME data, both within TMEs with respect to sub-sampling and between independent TME cores, and second, to compare the variability of TME data with that of field tests. A literature database with regard to semi-field and field tests should be established based on already published data (Jänsch et al. 2006).

4) Experimental setup, sampling, and analysis:

   a) Literature on microlysimeter experiments should be investigated for information on how excavation may affect the soil hydrology. The bottom of the soil column may be in contact with belowground soil or closed by a water-permeable inert material such as looped metal plates or porous ceramics. The different column closures will directly influence soil hydrology and therefore should be compared.

   b) Research and guidance is needed with respect to the minimum soil core size and to the appropriate sizes of soil populations in the cores that should be applied in TME studies. A minimum time period for

TME to equilibrate after removal from the field has not been defined yet. Neither is it known under which conditions the cores should be kept during the equilibration period regarding watering in case of dry periods, or covering them to protect from heavy rainfall if the cores are installed in the field.

c)   For persistent pesticides, the annual doses need to be added on top of the accumulation plateau that is formed due to previous applications. However, it is not clear which mode of application would be most appropriate: The test substance may be applied at concentrations resembling both the accumulation plateau concentration and the annual dose simultaneously, or the annual dose may be added after application of the accumulation plateau concentration and a corresponding aging period. Tests should be performed to compare both types of application.

d)   More research is needed to evaluate what statistical power is achievable in TME experiments. It should be investigated whether sub-sampling of a soil core at the defined sampling intervals will affect the soil communities of the residual soil columns compared to sacrificing individual soil cores at each sampling interval. Sub-sampling will considerably reduce the number of soil cores to be established for testing pesticide side effects on soil communities, and therefore will enhance the practicability of semi-field tests. In addition, research is needed to distinguish whether techniques like principal response curves are sufficient to explore community level effects, or whether diversity indices can play an additional role. This has been done for the aquatic compartment (Van den Brink and Ter Braak 1998) where the PRC approach was considered to be most appropriate.

e)   If recovery of the organisms from pesticide effects after crop harvest is not achieved, it may be necessary to extend the duration of the test over the next season or the next year. So far, TME studies have been performed for up to a period of 1 year: It should be tested whether the studies can be set up for even longer than 1 year to follow the long-term recovery of soil communities, if necessary. Considering the properties of TMEs, in particular the limited size, other methods might be more appropriate when studying recovery (e.g., field studies). From a regulatory point of view, a need exists for suitable triggers for conducting semi-field studies, including TMEs and guidance on summarizing and evaluating results from semi-field methods, including TME studies.

1 Introduction

1.1 SCOPE OF THE WORKSHOP

The scope of this guidance from the SETAC workshop PERAS ("Semi-Field Methods for the Environmental Risk Assessment of Pesticides in Soil") is to identify suitable semi-field test methods that are able to detect potential effects of pesticides on soil communities. Semi-field methods are defined as controlled, reproducible test systems that simulate the processes and interactions of components in a portion of the terrestrial environment, either in the laboratory, in the field, or somewhere in between.

The first mention of the potential for use of terrestrial model ecosystems (TMEs) in European pesticide risk assessment came in the European and Mediterranean Plant Protection Organization (EPPO) risk assessment scheme for soil organisms and functions in 2000, and also in the "Guidance Document on Terrestrial Ecotoxicology" under Council Directive 91/414/EEC (SANCO/10329/2002) (EC 2002), where TMEs were proposed as a potential higher-tier refinement step. How the method will "fit" into the current tiered risk assessment scheme was still unclear. However, semi-field tests may gain importance with the upcoming Regulation 1107/2009 (EC 2009), replacing Directive 91/414/EEC, particularly once questions are resolved regarding whether the soil risk assessment should focus on soil structure (i.e., community structure and biodiversity) or soil function (e.g., microbial respiration, organic matter breakdown).

The PERAS workshop organizing committee agreed that while TMEs might prove to be 1 suitable method within a tiered testing procedure (fitting between lab and full-scale field trials), the workshop should not focus solely on these to the exclusion of other methods that might be equally valid. TMEs should not be sold as a *fait accompli*, as different methods might suit different purposes. However, because of extensive experience with this method, there was more technical information available for discussion than for other semi-field and field methods.

The workshop was held as a SETAC–Europe workshop on October 8–10, 2007, in Portugal (Coimbra); the program is summarized in Appendix 1. Aims of the PERAS workshop were these:

- To highlight the current state of knowledge regarding semi-field methods and to identify the most appropriate methods to assess the impact of chemicals on soil community structure and function (see Chapters 2 and 3).
- To give a particular focus on higher-tier laboratory and semi-field methods that may be employed between first-tier laboratory tests and full-scale field studies. Special attention was paid to TME study types (see Chapter 4).
- To discuss technical aspects of the TME method in order to agree, as far as possible, on a standardized test method (see Chapter 5).

- To identify key gaps in knowledge and areas for further research and development in testing the effects of PPPs in soil (see Chapter 6).

After the first session with several keynote presentations addressing the scientific and regulatory background of higher-tier soil risk assessment, participants discussed specific topics in 3 breakout sessions. At each breakout session, 3 groups discussed predefined topics in parallel. Group rapporteurs presented the outcome of the group meetings to the plenary for further discussion.

After the workshop, the organizing committee compiled a draft document summarizing the presentations and the sometimes diverse discussions at the workshop. It was the editors' understanding from the workshop that there was general agreement for an outline TME method to also be included. While a draft method is shown in Appendix 2 for information, this does not represent an endorsement by the workshop that the TME approach is the only method of choice for higher-tier risk assessment. Following the commenting round and the inclusion of some changes and recommendations by participants, further concerns by a few participants who still felt that the document did not properly reflect the range of opinions expressed were again addressed. Having made some further changes, the editors are content that this document presents a fair summary of the PERAS workshop, which took place during a period of still rapid development in this area of risk assessment.

In order to provide an updated and holistic picture of recent relevant developments, the organizing committee also included some new additional information on semi-field testing that has become available since the workshop. It should be noted that while this information is of interest in the continuing debate concerning semi-field studies and their potential regulatory use, it was not endorsed by the workshop as a whole. This additional information is contained in the following references: De Jong et al. (2008, 2009), Kools et al. (2009), Scott-Fordsmand et al. (2008), Scholz-Starke et al. (2008), Theißen (2009), Van der Linden et al. (2008a, 2008b).

1.2   BACKGROUND

One of the many factors that may lead to spatial and temporal changes in soil biological communities is the use of pesticides in the agricultural landscape. Several tests methods in a tiered approach are available to characterize the potential impact of xenobiotics in soil. Semi-field methods are defined as controlled, reproducible test systems that attempt to simulate the processes and interactions of components in a portion of the terrestrial environment, either in the laboratory, in the field, or somewhere in between. According to the discussions and conclusions in the workshop, methods proposed so far can be classified in 3 main and 6 subgroups (Figure 1.1), depending mainly on the treatment of the soil (artificially assembled systems vs. intact soil cores) and the environmental conditions (controlled vs. field). A detailed discussion on this classification is given in Chapter 4.

The potential for the use of TMEs, or other semi-field study designs, may gain importance with the upcoming Regulation 1107/2009 (EC 2009), replacing Directive 91/414/EEC, in which the focus of soil risk assessment is both on the structure (biodiversity, in particular community structure) and function (e.g., microbial respiration,

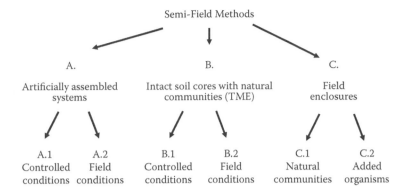

FIGURE 1.1 Proposal for the classification of terrestrial semi-field methods (for further information see Section 4.2).

litter breakdown) of the soil community (Morgan and Knacker 1994; EFSA 2007). TMEs may also fit into the proposed Dutch decision tree for persistent pesticides as a method for higher-tier assessment (van der Linden et al. 2008b).

Only few validated higher-tier laboratory or semi-field methods are available to assess structural and functional effects of pesticides in soil. In this context, the SETAC Europe workshop PERAS was organized. Aiming to present and discuss the state of the art with a focus on semi-field methods such as terrestrial model ecosystems (TMEs), 52 experts from academia, industry, and authorities, e.g., the European Food Safety Authority (EFSA), Organization for Economic Cooperation and Development (OECD), and national pesticide registration agencies, were invited from Europe, Brazil, and the United States.

Although these PERAS proceedings focus mainly on the use of semi-field methods for the risk assessment of plant protection products (PPPs), they can also be useful for other groups of potentially toxic substances, such as industrial chemicals or veterinary pharmaceuticals (e.g., Boleas et al. 2005). However, it should be noted that the exposure design for substances other than PPPs may differ. While semi-field methods have already been used for the assessment of material like fly ash (Van Voris et al. 1985) or contaminated soils (Kools 2006), there is currently not enough experience to recommend them for site-specific risk assessment.

In contrast, methodologically the PERAS proceedings focus on class B tests, i.e., intact soil cores with natural communities (TMEs), since they have regularly been applied in ecotoxicological studies. TMEs may be used for the environmental risk assessment of industrial chemicals, biocides, and plant protection products (Weyers et al. 2004; EMEA 2007). The potential for the use of TMEs in pesticide risk assessment was mentioned in the EPPO risk assessment scheme for soil organisms and functions in 2000 and also in the "Guidance Document on Terrestrial Ecotoxicology" under Council Directive 91/414/EEC (SANCO/10329/2002) (EC 2002). While TMEs were mentioned as a potential higher-tier refinement step, it is not clear precisely how such methods would fit into a tiered risk assessment scheme. Therefore, this aspect was one of the topics discussed during the workshop.

2 Ecological Considerations

2.1 SCOPE

In the following, some ecological background information of use in understanding the possibilities and constraints of terrestrial semi-field methods is presented.

Soil forms a thin layer over the earth's surface and acts as the interface between the atmosphere and lithosphere. Soil consists of mineral material, organic matter at various stages of decay, plant roots, soil biota, water, and gases. On the one hand, soil provides a medium for an astounding variety of organisms that use the soil as habitat and a source of energy. On the other hand, the organisms contribute to the formation of soil by influencing the soil's physical and chemical properties and the nature of vegetation that grows on it. The 5 interacting soil-forming factors are the parent material, climate, relief, biota, and time. Natural and anthropogenic factors lead to spatial and temporal changes in soil biological communities.

The definition of soil includes soil organisms as an integral part of soils. The term "living matter" can be specified as "organisms" or, even better, "plants, microorganisms, animals and their interactions (including functions)," as stated, for example, in the German plant protection law (PflSchG 1998). This is not a trivial statement since soil quality is often defined in a strictly anthropogenic way, meaning that it is seen as a substrate that "sustains plant and animal productivity, maintains or enhances water and air quality, and supports human health and habitation" (Karlen et al. 1997, p 6). This definition does not highlight that soil is also a habitat for organisms that are crucial for many soil functions. From a biological point of view, soil is defined as the uppermost mineral layer, usually down to 1 m; however, the predominant number and biomass of invertebrates is located within the uppermost 10 to 20 cm in most soils, along with the soil surface. Part of the soil surface is the litter layer, which is an integral part of the soil, not only in many forests but also in grasslands and some permanent crops.

Soils and thus also the soil organism communities living in them differ considerably over different spatial scales, from centimeters to ecoregions. In the context of the PERAS proceedings the information provided covers mainly temperate regions of the northern hemisphere, partly due to the fact that knowledge of soils and their communities is lacking in other parts of the world. However, the main reason is that outside of the European Union (and, to a lesser extent, in a few other countries like Canada) there is no (potential) legal requirement for soil risk assessment (including semi-field or field methods) in regulatory ecotoxicology. Concerning different land use forms, the focal point of the PERAS proceedings is the agricultural landscape, comprising crops, meadows, orchards, grassy field margins, hedges, floodplains, etc. Whether only the in-crop soil community needs to be considered, or also off-crop soils, remains to be discussed. However, some comparative research on off-crop soils may be useful to determine the impact of various anthropogenic inputs.

The term "soil organisms" covers a wide range of microbes, plants, and invertebrates, which spend an important part of their life cycle (e.g., as larvae) or their whole life within the soil (including the leaf litter layer). For reasons of practicality, the few vertebrates that fulfill this definition will not be handled further. In addition, plants have played only a limited role in semi-field methods so far, meaning that besides microbes, soil invertebrates are the main group of interest in the context of terrestrial semi-field methods.

Soil invertebrates are usually classified according to their size (either length or diameter) and to the trophic level they belong to. One common classification scheme, originally proposed by Swift et al. (1979), and later modified by Beck (1993), is given in Figure 2.1. Some typical soil organisms are compiled in Figure 2.2.

2.2 SHORT OVERVIEW OF "TYPICAL" SOIL ORGANISM COMMUNITIES

Understanding the complex patterns of soil biodiversity and the factors that control them is the main focus of soil community ecology. Information on the actual diversity of groups of belowground soil biota is very sparse compared to that of aboveground organisms, especially at the species level. This lack of knowledge is understandable: Because soil organisms are not easily seen, they are very difficult to study and lack the "sentimental appeal" that many aboveground species have.

When discussing soil organism communities, it is important to distinguish between the structure and the function of the soil biocoenosis or community, which can be defined as follows (Schäfer and Tischler 1983; Odum 1985):

Structure: Composition of the soil biocoenosis (biodiversity), described at the species level (i.e., abundance, biomass, diversity, and dominance).
Function: Biologically determined processes in the soil ecosystems, based on the interaction of its different components (i.e., nutrient cycling, community respiration, or most prominently, organic matter breakdown).

In the following, different aspects of the structure and functions of soil organism communities will be discussed. For more ecological details see Bardgett (2007), and for consequences in monitoring programs see Römbke and Breure (2005).

2.2.1 THE STRUCTURE OF THE SOIL BIOCOENOSIS

Soil biota are thought to harbor a large part of the world's biodiversity and to govern processes that are regarded as globally important components in the cycling of organic matter, energy, and nutrients (e.g., Griffiths et al. 2000). Rough estimates of soil biodiversity indicate several thousand invertebrate species apart from the largely unknown microbial and protozoan diversity (e.g., 1500 to 1800 invertebrate species in a German beech forest (Weidemann 1986)). By far the most dominant groups of

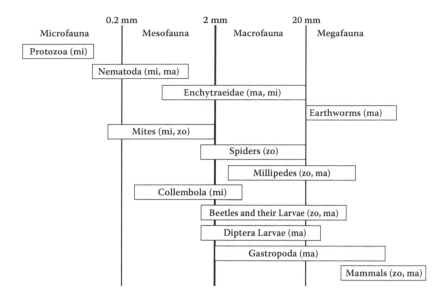

FIGURE 2.1   Body size of common soil invertebrate groups, including their assignment to certain trophic levels (ma = macrosapro- and macrophytophagous, mi = microphytophagous, zo = zoophagous, including necrophagous). (Modified from Beck, *Biologie in unserer Zeit*, 23, 286–294, 1993.)

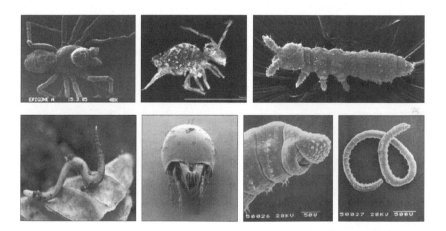

FIGURE 2.2   Common soil invertebrates (from the top left, clockwise): Araneae, Collembola, Enchytraeidae (whole body and head), Oribatida, and Lumbricidae (photos by Hubert Höfer, Franz Horak, David Russel, Jörg Römbke, Andreas Toschki, and John Jensen).

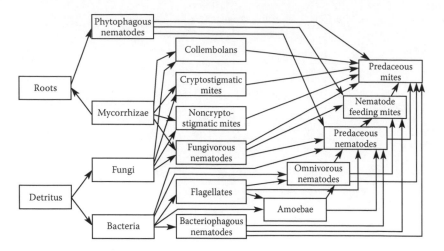

FIGURE 2.3 Structure of parts of the soil food web. (Adapted from Bardgett, *The Biology of Soil*, Oxford University Press, Oxford (UK), 2007.)

soil organisms, in terms of numbers and biomass, are the microbial organisms, i.e., bacteria and fungi. Estimates on the number of microbial species (genotypes) in the soil range from 10^4 to 10^5 per gram (Torsvik et al. 1990; Dykhuizen 1998). Besides these organisms, soil ecosystems generally contain a large variety of animals, like protozoa (bacterivores, omnivores, predators), nematodes (bacterivores, fungivores, omnivores, herbivores, predators), microarthropods such as mites (bacterivores, fungivores, predators) and Collembola (fungivores, predators), and enchytraeids and earthworms (both being mainly saprophagous). In addition, a high number of macrofauna species (mainly arthropods like beetles, spiders, diplopods, and chilopods, as well as snails) live in the uppermost soil layers, the soil surface, and the litter layer.

These organisms are organized at different trophic levels and groups within the soil food web (Figure 2.3). It should be noted that in this example several common groups of soil organisms, such as earthworms or enchytraeids, are missing for reasons of clarity.

2.2.2 Functions of Soil Organisms

With its large diversity and complexity, the soil community has a strong impact on soil processes and the way in which these processes may vary in time and space. Most noteworthy are

- decomposition of organic matter, thus supporting the cycling of nutrients;
- fixation of nitrogen from the air, making it available for plants;
- stabilization of soil aggregates, specifically by building clay-humus complexes;
- improvement of soil porosity due to burrowing activities;
- degradation of anthropogenic compounds like pesticides;
- influencing of soil pH by nitrification and denitrification, resulting in mobility changes of heavy metals;

- influencing of heavy metal mobility under different redox conditions (e.g., in the sulfur cycle, especially important in areas with fluctuating water tables); and
- being prey for many aboveground organisms.

As these processes also determine nutrient availability for uptake by plants, the belowground decomposer food web interactions also influence aboveground primary productivity and carbon sequestration (Scheu 2001; Tscharntke and Hawkins 2001). In fact, plant productivity appears to increase in response to a reduced turnover of the microbial biomass due to stabilized carbon content and soil pH. Vascular plants are known to be extremely sensitive to microbial symbionts. Mutualisms in new environments are key functions for competitiveness (and successful migration due to human activities and global climate change) of most plants, since microbial symbionts are required to induce N-fixation.

Soil organisms are assumed to be directly responsible for soil ecosystem processes, especially the decomposition of soil organic matter and the cycling of nutrients (Bardgett and Chan 1999). For example, the soil biomass is known to process over 100 000 kg of fresh organic material each year per hectare (25 cm top soil layer) in many agricultural systems. This processing includes the decomposition of dead organic matter by microbes as well as the consumption and production rates in the soil community food web (Hunt et al. 1987; De Ruiter et al. 1993, 1998). The soil food web is defined as the structure and interactions across and between the communities of soil-living organisms, which are linked by conversions of energy and nutrients as one organism eats another. Therefore, most food web models merely provide a way to connect the dynamics of populations to the dynamics in ecological pathways within the cycling of matter, energy, and nutrients (Pignatti 1994). The food web models have to use the observed abundance, biomass, or energy of the various groups of soil organisms as input variables. By incorporating the multivariate habitat response relationships for the occurrence of organisms, they aim at assessing the effects of (changes in) abiotic and management conditions in scenario studies. Although this approach enables food web models to provide a reliable way to analyze the dynamics of soil populations in the context of soil community structure as a whole, these models do not unravel the key problem of sustainability as an effect of environmental or agronomic change with respect to stability and biodiversity. From this point of view, the most important remaining problem is the definition of a reference system.

2.2.3 THE INTERACTION BETWEEN STRUCTURE AND FUNCTION

Indicators of functional biodiversity could best be based on the measurement of processes. However, soil processes fluctuate strongly in time and space. Establishing a mean annual value of a process requires an intensive sampling program, and is difficult to establish on a national scale. Therefore, it is more practical to use the species composition, aggregated in functional groups, as an indicator for processes. The relationship between species composition and ecosystem functioning is difficult to quantify. When species disappear, others may become more dominant and take over a link in the overall process. It is possible that a process will continue while species composition has changed or degraded (this is so-called "functional redundancy"). So, the preservation of biodiversity cannot be guaranteed by measuring process values. Processes are too general or insensitive as an early warning indicator.

When discussing the species composition in soil, one has to remember that due to the extreme heterogeneity (both spatially and temporally) of soil, the number of species is often very high (Bardgett 2002). Therefore, it is very difficult to identify simple and predictable relationships between the diversity and function of soils. Soil functions seem to depend mainly on the complexity of biotic interactions (e.g., between the components of the food web). However, in some cases individual species (often defined as key species or ecosystem engineers) can dominate the function of the whole system, for example, large anecic earthworms in many nonacidic soils (Lavelle et al. 1997) or enchytraeid worms in acid forest soils (Laakso and Setälä 1999). Thus, it is important to distinguish between species when assessing the habitat function of the soil, meaning that the species level is the most accurate taxonomic level when using natural communities for bioindication purposes, although it requires a high amount of labor, knowledge, and time (Nahmani et al. 2006).

In general, heavy pollution or disturbances select for a few resistant (tolerant) species. In such a situation the ecological basis for processes has become very narrow. When the resistant species disappear, a process stops and the specific function is permanently affected. The indicator system for functions of the soil organism community is based on the following approach: the threat of vital soil processes can be expressed by comparing the number of species in functional groups of a certain area with its reference (undisturbed locations). A process is assumed to continue to exist with fewer species, in which case the risk of instability and uncontrolled fluctuations will increase.

The assessment of any actual interaction between structure and function in the field suffers for the lack of a reference choice (Beck et al. 2005). The use of the landscape by humans has brought profound changes aboveground and belowground, and even provoked global climate changes. In this context, knowledge of soil structures and functional processes becomes increasingly relevant. If one adds the food web perspective to the previous characters, one will obtain the most immediately perceptible, sensitive elements for the characterization of a (soil) community: species composition, food web structure, and individual phenology. Soil communities are complex, open systems with exchanges of matter and energy controlled by feedback loops (cf. Odum 1971). The presence of these loops enables the ecosystem to withstand unpredictable reactions to environmental stress. Dynamic models may provide unique possibilities to assess the ecological risk aside from our perception (i.e., "expert judgment") due to a categorization of ecological processes aside from a static reference.

Although in Central Europe plenty of attention has been given to the human influences on the aboveground vegetation recently (e.g., Mucina et al. 1993; Zechmeister and Moser 2001) as well as during the last 2500 years (long-term studies by Vera 2000), still very little is known about human-induced cascade effects in belowground systems. The ruderal species carefully described by these authors for the upper vegetation still have no analogous taxa for the soil. This confirms a tremendous lack of knowledge of soil organisms, albeit sufficient evidence of environmental impact on the composition and abundance of soil biota at different trophic levels exists (Wall et al. 2001), and confirms the extent to which species are "redundant." Rapport et al. (1998) suggest resilience (measured in terms of a system's capacity to maintain both structure and functions under environmental stress), organization (diversity of interactions within a system, mainly the loops described before), and vigor (productivity) as indicators of ecosystem quality. Also, their approach requires a careful selected reference for any local scale assessment, since the concepts of resilience, structure, and vigor (processes) rely entirely upon dynamic trends.

2.3   QUANTITATIVE DESCRIPTION OF SOIL ORGANISM COMMUNITIES

Despite several decades of soil biological studies, it is still very difficult to provide average abundance and biomass values for soil invertebrates. On the one hand, this is caused by the high variability in time and space as well as differences in sampling methods used. On the other hand, most work has been performed in forest soils of temperate regions, while other ecoregions, like the tropics or land use forms, have seriously been neglected—which is in particular true for crop sites. Finally, due to difficulties in sampling as well as in taxonomy, mesofauna groups were studied much less than macrofauna groups, especially earthworms. In the following, numbers indicate the range of abundance of several organism groups (Table 2.1). However, when considering the low number of studies and their often low comparability, these numbers are just a rough indication. Using species number, sampling efficiency, and ecological relevance as main criteria, the following organism groups are usually recommended for semi-field studies, both in artificially assembled systems and in terrestrial model ecosystems (groups A and B):

* Macrofauna: Earthworms (rarely isopods)
* Mesofauna: Springtails, mites, and enchytraeids

However, the mean numbers in Table 2.1 are strongly biased by the fact that most of these studies were performed in forest soils, usually without anthropogenic stress but with high humus content. The maximum numbers are based on optimum conditions (e.g., the high numbers of enchytraeids were found in an acid moor soil where almost no other invertebrates could occur (Peachey 1963)). In agricultural soils, which are characterized by several factors stressing soil invertebrates (e.g., plowing, fertilizers, compaction, and pesticides), these numbers are considerably lower. As an example, abundance, biomass, and species numbers for some selected micro- and mesofauna groups from agricultural soils of Central Europe are given in Table 2.2.

Since such numbers are not available in detail for most of the mesofauna groups, only earthworms and collembolans will be discussed in the following. An overview

TABLE 2.1

Abundance of the most important soil invertebrate groups in temperate regions (mainly forests); average and maximum values

Size class	Organism group	Mean ind/m²	Maximum ind/m²
Microfauna	Flagellata	100,000,000	10,000,000,000
	Nematoda	1,000,000	100,000,000
Mesofauna	Acari (mites)	70,000	400,000
	Collembola	50,000	500,000
	Enchytraeidae	30,000	300,000
Macrofauna	Lumbricidae	100	500
	Gastropoda	50	1000
	Isopoda	30	200
	Diplopoda	100	500
	Beetles (larvae)	100	600
	Diptera (larve)	100	1000

Source: Modified from Dunger, *Tiere im Boden*, A. Ziemsen-Verlag, Wittenberg (DE), 1983.

on the number and biomass of earthworms at temperate agricultural sites is given in Table 2.3. It seems that the variability of the numbers given in this table is very high, but it has to be kept in mind that the number of sites (e.g., orchards) was small and that the individual sites were quite heterogeneous. However, when just 1 number should be given concerning the "average" earthworm population size, a density of 80 ind/m², a biomass of 5 g DW/m², and a species number of 4 could be given. In a comparison of the earthworm community of crop plots and their respective field boundaries of 5 South German sites, clear differences (based on various sampling dates) were found (Ehrmann 1996):

- abundance and biomass were significantly lower in the cropped areas (5% to 50%),
- the number of species was also lower (3.4 vs. 4.6), and
- on the crop plots the percentage of endogeics was higher and that of adults was lower.

The numbers given in this table vary considerably. As an example, the comprehensive study of earthworm numbers, biomass, and species composition performed at 102 agricultural and 20 grassland sites all over Bavaria is worth mentioning (Bauchhenss 1982, 1997). In total, 38,000 worms were collected during 2 sampling series between 1985 and 1995. Since this data set is so much larger than all other comparable studies, it is probable that most of the conclusions drawn from this data set can be used for other regions of Germany and Central Europe:

TABLE 2.2
Abundance, biomass, and species number, given as range of mean numbers of selected micro- and mesofauna groups in agricultural soils of central Europe

Organism group	Abundance (ind/m²)	Biomass (mg DW/m²)	Species number
Nematoda[a]	3000–13,000	≈440	17–20[b]
Acari (mites)	<1000–5000	≈120	3–10
Collembola	1500–33,000	≈120	17–38
Enchytraeidae	2000–30,000	110–640	3–22

Note: ≈ indicates numbers deduced from grassland sites. Based on literature reviews (Petersen and Luxton 1982; Römbke et al. 1997) and monitoring studies (Filser 1995; Römbke et al. 2002; Theißen 2009).

[a] Numbers given per kg soil DW (dry weight).

[b] Families, not species.

TABLE 2.3
Compilation of data describing the "typical" earthworm community of agricultural sites in central Europe (or parts of it)

Area and habitat	Number (ind/m²)	Biomass (g DW/m²)	Number of species	Reference
Central European crop sites	6–453	0.5–15.2	0–11	Römbke et al. (1997)*
Central European crop sites	74.7	4.8	3.6	Römbke et al. (1997)**
	0.9–187	0.1–12.1	1–7	
Switzerland crop sites	—	11.0***	7.0	Stähli et al. (1997)
		0.6–28.6	5–9	
Bavarian crop sites	9	0.6	3.0	Bauchhenss (1997)
	0–280	0–?	0–?	
Germany vineyards	36.7	6.8	1.9	Kühle (1986)
	0–83	0–20.7	0–4	

Note: When possible, ranges are also given: *, including the compilations of Satchell (1983) and Lee (1985); **, excluding UK; ***, recalculated from fresh weight; —, no literature data available. The UK sites were excluded since they were very often located in former peat bog areas with acid soil.

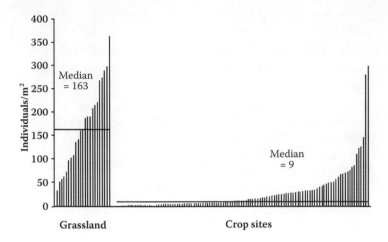

FIGURE 2.4   Number of earthworms (ind/m²) found at 116 crop and 20 grassland sites in Bavaria. (Adapted from Bauchhenss, *Schriftenr Bayer Landesamt Umweltschutz*, 6/97, 222, 1997.)

- The abundance and biomass of earthworms are significantly higher at grassland sites than at crop sites (median values: 163 to 9 ind/m² and 104 to 6 g/m²).
- The abundance at grassland sites varies between 40 and 360 ind/m² (one order of magnitude), and at crop sites between 0 and 280 ind/m² (several orders of magnitude) (Figure 2.4).
- The individual body weight of each species is higher at crop sites than at grassland sites (Figure 2.5). This observation may reflect the less optimal food situation at crop sites: The worms grow large since the resources are just enough to survive and to grow very slowly, but not enough to reproduce—a behavior especially well known for *Lumbricus terrestris* (Graff 1953).

In general, abundance and biomass of earthworm populations are much higher in pastures than in crop sites due to better food supply, better buffering of the soil against climatic events, and no, or only small, anthropogenic impacts (Edwards and Bohlen 1996). This is true for general comparisons as well as in cases where the 2 sites are located close to each other (Decaens et al. 2002; Römbke et al. 2002).

Species numbers and abundances of the collembolan coenosis within Central European crop sites are displayed in Table 2.4. Numbers vary depending on, e.g., different site characteristics or sampling intensity. Averaging the data results in ~12,000 ind/m² and ~23 species per site. It has to be kept in mind that the species structure can differ considerably between sites. A literature review screening the data of 12 authors and 29 data sets in crop sites of Central Europe lists in total 122 different collembolan species (Theißen 2009).

In comparison with data from other openland habitats in Central Europe, the average abundance is in an upper mean range, while the average species number is

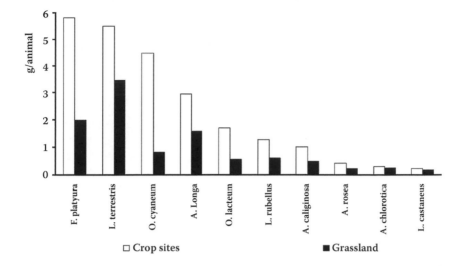

FIGURE 2.5 Individual biomass of earthworms (g fresh weight (FW)/animal) found at 116 crop and 20 grassland sites in Bavaria. (Adapted from Bauchhenss, *Schriftenr Bayer Landesamt Umweltschutz*, 6/97, 222, 1997.)

TABLE 2.4

Compilation of data describing the "typical" Collembolan community of agricultural sites in central Europe, assessed by different soil extraction methods

Area and habitat	Number (ind/m²)	Number of species	Reference
Poland, crop sites (sugar beet)	8660–21,810	17–23	Czarnecki and Losinski (1985)
Germany, crop sites (rapeseed/hop)	2500–7500	17–19	Filser (1992)
Germany, crop sites (barley/wheat)	23,000	38	Heimann-Detlefsen (1991)
Germany, crop sites (oat/wheat/maize)	16,324–33,305	26–29	Lübben (1991)
Poland, crop sites (tomato/barley)	1505–5947	17–27	Sterzynska (1990)
Germany, crop sites	6429–13,404	23–30	Ufer (1993)

low. According to different citations in pastures, mesophilic and moist meadows, and grassy field margins, average collembolan densities of 9900, 8500, 36,300, and 2700 ind/m² and average species numbers of 36, 33, 35, and 44 species per site can be expected (Theißen 2009).

2.3.1 HORIZONTAL DISTRIBUTION

Earthworms are not evenly distributed in soil—a fact that has been known at least since the late 1940s. Probably the uneven distribution of physicochemical soil properties, but mainly of food, as well as the reproductive potential and dispersive powers of the individual species, and last but not least, historical events (e.g., disturbance) are responsible for this pattern (Edwards and Bohlen 1996). Satchell (1955) found clearly higher aggregation rates for juveniles than for adults in an English grassland. Experimentally, evidence for aggregation was due to soil moisture and, more often, to food resources (e.g., higher numbers of *Dendrobaena octaedra* and *Lumbricus rubellus* beneath dung pats in pastures (Boyd 1958)). However, there are examples where individuals of *Aporrectodea rosea* and *Lumbricus castaneus* were highly aggregated in a homogenous pasture: Probably the worms reproduce more quickly than they can disperse from the breeding site (Satchell 1955). Surprisingly, the horizontal distribution of earthworms has rarely been studied at crop sites (probably because these sites have been considered homogenous). In a French maize field, earthworm abundance was higher within than between rows by a factor of 2, which can be explained by migration of adult worms into the corn plots, attracted by the microorganisms living close to the roots or by root exudates themselves (Binet et al. 1997). Another possibility is that they were driven out by heavy machinery, which leads to a compaction of the soil between rows.

Despite the fact that springtails are small, they are often able to move long distances. This is mainly true for the epigeic and hemiedaphic species (those living on or close to the soil surface, including litter), while euedaphic species (those living in the mineral soil) have the tendency to be more sedentary. Like earthworms, collembolans are also not evenly distributed in soil. Studies from Usher (1969, in Hopkin 1997) show that the main portion of springtail species show an aggregated distribution rather than a random or uniform distribution. As main reasons for forming aggregations, ideal moisture, suitable pore sizes, and food sources are cited (Hopkin 1997). Even in seemingly similar contiguous arable fields with similar histories, the species composition of Collembola varies considerably, which also influences the effects of pesticides applied at these sites (Frampton 1999).

2.3.2 VERTICAL DISTRIBUTION

Different species of lumbricids inhabit different depth zones in the soil, depending mainly on climatic conditions and the availability of food, thus considerably changing seasonally (Edwards and Bohlen 1996). As indicated by the Ecological Considerations workgroup, epigeics live close to the soil surface, while endogeics inhabit the uppermost 15 cm (adults of *Octolasion cyaneum* go down to 56 cm).

Among anecics (vertical burrowers), *Aporrectodea longa* prefers the uppermost 45 cm and *Lumbricus terrestris* can go down to 2.5 m but is usually restricted to a depth of about 1 m (Edwards and Bohlen 1996). These anecics stay active as long as possible by retreating to the bottom of their burrows during extremes of heat or cold. In addition, it is well known that endogeic and epigeic juveniles tend to feed more closely to the soil surface, while the adults show the "typical" behavior of their respective ecological group (Briones and Bol 2003). Gerard (1967) showed that the vertical distribution of 6 common earthworm species (*Allolobophora chlorotica, Aporrectodea caliginosa, Aporrectodea longa, Aporrectodea nocturna, Aporrectodea rosea, Lumbricus terrestris*) in England mainly depends on the moisture and temperature of the uppermost soil layers. Nearly all cocoons of these species were found in the uppermost 15 cm, most being in the top 7.5 cm. The same observations were made in Sweden (Rundgren 1975) and Germany (Peters 1984).

The vertical niche differentiation of collembolans is correlated along with species-specific morphological traits. According to the "life form concept" (after Gisin 1943; Christiansen 1964) springtails can be categorized based on the size of furca (springing organ) and antennae, the number of ocellae, and their pigmentation into epigeic, hemiedaphic, and euedaphic species. Although some species are strictly confined to a certain soil layer, many species have a broader vertical niche. Since they do not have the ability to create burrows, springtails depend on the existing pore system and burrows made by, e.g., earthworms. The highest density of collembolans in openland habitats of Central Europe can be expected in the upper 5 to 10 cm soil layer. Vertical migration regularly exists and is mainly induced by climatic factors.

2.3.3 VARIABILITY IN TIME

Earthworm populations are dynamic, with constant changes in terms of size, biomass, and dominance spectrum due to seasonal variations, direct perturbations, or a combination of interacting factors (Bembridge et al. 1998). Unfortunately, when looking at literature data it has not always been possible to distinguish between earthworm numbers and earthworm activity: In the latter case only electrical or chemical expellant methods were used (Edwards and Bohlen 1996), while in the former a combination of hand sorting and formalin extraction was usually performed. In general, the population size of earthworms in temperate regions follows the seasonal cycle: high numbers in spring and autumn and low numbers in summer (due to drought) and winter (due to coldness) (Evans and Guild 1947; Gerard 1967). In addition to such seasonal changes, many earthworm species are distinctly diurnal in their activity; e.g., *Lumbricus terrestris* is usually active between 6 p.m. and 6 a.m., which is intrinsic, i.e., partly independent of temperature and light triggers (Edwards and Bohlen 1996). In a 10-year study, the seasonal variability of earthworm abundance and biomass was studied in 2 small plots at an English grassland site (Bembridge et al. 1998). Large fluctuations were found, leading to overall ranges of 72 to 512 ind/m² and 35 to 253 g FW/m² (fresh weight) at individual sampling dates (spring or autumn, respectively). While these ranges seem to be high, they are due to normal annual variations. In addition, the differences between the 2 plots were negligible.

As expected, climatic factors are probably responsible for most of the variability, which became obvious in 1983 when the numbers decreased strongly after a severe drought period.

Since soil moisture and temperature are the 2 most important ecological factors determining the presence of soil collembolans, the variability of abundance in time is linked with seasonal and diurnal climate changes (Frampton et al. 2000). But it seems to be widely accepted that seasonality is a consequence of the climatic conditions (Christiansen 1964; Hopkin 1997). The occurrence of species can differ both within season and from year to year (Frampton 2001; Frampton et al. 2001). Since the average summer in Central Europe is warm and dry and the average winter cold and humid, peaks of collembolan abundance are regularly expected between autumn and spring.

2.4   POTENTIAL EFFECTS OF PESTICIDES ON SOIL ORGANISM COMMUNITIES

Again, earthworms are taken as an example because they are by far the best studied organism group. Due to their simplicity in testing and ecological relevance, the compost worm *Eisenia fetida* was selected as the first ecotoxicological test species for the soil compartment (OECD 1984). More importantly, the first internationally standardized terrestrial field study with soil organisms is the earthworm field test (ISO 1998). In addition, earthworms have been used successfully both in artificially assembled systems and in terrestrial model ecosystems (TMEs) (Burrows and Edwards 2004; Römbke et al. 2004).

In general, earthworm species do not differ strongly in sensitivity toward many PPPs (i.e., there is no most or least sensitive species; Heimbach 1985), but of course certain species can react differently to selected compounds (Bauer and Römbke 1997). When comparing results of laboratory tests with *Eisenia fetida* with those from other species, differences in LC50 values were nearly exclusively within a factor of 10. One noteworthy exception is propoxur, where the LC50 of *Eisenia fetida* was 72 times higher than that for *Aporrectodea longa* (Jones and Hart 1998). However, some species are more exposed to chemicals due to their behavior (Edwards 1983). In particular, *Lumbricus terrestris* can be affected in several ways: by direct contact with applied products when they are at the soil surface, feeding on contaminated leaves, contact with contaminated soil, exposure via soil pore water, or exposure to aqueous PPP solutions being washed into their burrows (Edwards et al. 1995). Among older plant protection products (i.e., those marketed before ca. 1980) are many compounds (e.g., lindane or dieldrin) showing a high acute toxicity for earthworms as well as persistence and even bioaccumulation. For example, carbofuran can be accumulated in earthworms to an extent that predators like buzzards are affected after feeding on lumbricids (especially important in the autumn plowing season; Dietrich et al. 1995). Due to such old and often very toxic and persistent compounds (including copper compounds used as fungicides), orchards treated with them may still have very low earthworm populations for many years (Edwards et al. 1995). Additionally, the combined effect of 2 compounds applied together (e.g., carbofuran and atrazine) is higher than addition of the effects observed after individual applications (Lardier and Schiavon 1989). Only for specific reasons (e.g., at golf courses) are vermicides used, e.g., organophosphates or, in the past, mercury chloride (Lee 1985).

Besides acute and chronic effects, some pesticides can also cause indirect effects. The best known examples of indirect effects on earthworms are caused by herbicides (Bembridge et al. 1998), which usually show a very low toxicity toward earthworms at the recommended application rate, thus being classified as "of no concern." In the field they can have positive effects on earthworms, since they temporarily improve the food supply due to dead organic matter (even mowing can increase the earthworm biomass; Todd et al. 1992). However, in the long run they cause a food shortage, which leads to a decrease of the abundance and biomass as well as a change in the dominance spectrum of earthworms. This reaction pattern is similar to the situation that can be observed in ailing forests, e.g., after being stressed by acid rain (Coderre et al. 1995). Another indirect effect caused by herbicides is a consequence of altering the extent of plant cover, and thus the microclimate of the topsoil (Curry 1998).

Site-specific factors like soil properties also have to be taken into consideration, such as the influence of soil type on the bioavailability of some PPPs. Bioavailability has been shown to be higher in sandy soils than in loamy soils (Lofs-Holmin 1982). Also, cases have been reported in which PPPs behave differently due to the action of earthworms: The degradation of atrazine differs in the middens of *Lumbricus terrestris* compared to the surrounding soil. Atrazine degradation is enhanced in middens, which can be traced to the higher levels of microbial activity, which in turn is probably related to the enriched carbon content of the middens (Akhouri et al. 1997). Another important factor is the mode of application: For example, Ruppel and Laughlin (1976) demonstrated that an "in furrow" application has a much lower influence on earthworm populations than other forms of band applications or even spraying. Frequent treatments with toxic and/or persistent compounds over long periods should ideally be avoided. For example, it has been demonstrated that the long-term use of copper fungicides drastically reduced the number of earthworms in an English orchard (Raw 1962), and the application of high doses of insecticides (mainly phorate) eliminated earthworms from grassland plots (Clements et al. 1991).

An example showing the consequences of long-term effects is the use of the fungicide benomyl, which was often applied in apple plantations where it could cause severe side effects on earthworms (especially the "key species" *Lumbricus terrestris*). Consequently, distinct and, in some cases, even long-lasting effects on litter degradation were observed (e.g., Kennel 1990). In the long term, the fungicide applications even proved to be counterproductive to the extent that leaves covered with spores of the target fungi were not consumed, and therefore the fungi were no longer inactivated by the earthworms. As a result, fungi grew in much higher numbers than prior to application of the fungicide. Comparable experiences have been found in hay meadows (Stockdill 1982; Hoogerkamp 1987). However, when benomyl was applied to an English grassland site once per year for 10 years, interesting results were seen (Bembridge et al. 1998). While the population decreased strongly after application of 5 kg benomyl/ha within the first 4 years of the study, the use of 2 kg benomyl/ha applied later on in the same study caused comparable effects for only a few years. In the last 3 years, no decrease was found at all. Such experiences might support the approach of using a broad range of different test species and working with adverse effect scenarios and hypotheses involving entire processes or functions rather than individual standard species. Alternatively, this uncertainty and variability in potential responses could be accounted for in risk assessments based upon more limited data sets.

The authors did not discuss this result, but the following reasons might be possible (see also Edwards and Brown 1982):

- Species sensitive to benomyl were extinct; so, other species were able to compensate their number and biomass due to the higher amount of food available.
- Due to unknown reasons, the control numbers were much lower in the second half of the study period, which makes the identification of differences difficult.
- Genetic or physiological adaptation to benomyl is unlikely due to the short period of time available (exposure occurred only for 5 years at most), and the same species-specific effect pattern was observed after each application of benomyl.

Compilations of results from laboratory tests as well as field studies with many commonly used PPPs can be found in Lee (1985), Edwards and Bohlen (1992, 1996), Högger (1994), and Jones and Hart (1998). Referring to these authors, the effects of the main use classes of PPPs can briefly be summarized as follows, based on the available literature data (which do not necessarily reflect data submitted to authorities):

- Among insecticides, acaricides, and nematicides, the majority of older pesticides are toxic to very toxic to earthworms (e.g., many organochlorines, organophosphates, and carbamates), while more relatively recent substances like pyrethroids (Inglesfield 1984) or insect pathogens (in particular *Bacillus thuringiensis*) usually do not harm earthworms at field-relevant concentrations (Beck et al. 2004).
- Among fungicides, while some compounds, like benomyl and carbendazim, are very toxic, others are not. Also of note are the moderately to highly toxic copper salts like copper oxychloride and copper sulfate, which have been strongly accumulated in many vineyard soils.
- Practically all fumigants are highly toxic to earthworms.
- No significant direct toxicity to earthworms could be detected for the investigated herbicides, but as discussed earlier, their indirect effects should also be considered.

An overview of the current knowledge on the effects of pesticides on soil invertebrates in general and earthworms and collembolans in particular is given by Frampton et al. (2006), summarizing the results of laboratory tests, and by Jaensch et al. (2006), for higher-tier (TME and field) studies. While it was possible to calculate species sensitivity distributions (SSDs) for some pesticides, the determination of no-observed-effect concentrations (NOECs) at the field level was severely hampered by the lack of available data. Concerning the study of effects of pesticides on earthworms and Collembola in TMEs and the field, the best example is a study with the fungicide carbendazim. This compound was applied in the formulation Derosal® to TMEs and the respective field plots at 4 European sites (Römbke et al. 2004). The sites selected had an earthworm coenosis representative of the different land use types and regions. In addition, the differences between the lumbricid coenosis found in the TME and the respective field sites were in general low. A high variability was found between the replicate samples, which reduces the probability of determining significant differences through statistical evaluation of the data. Similar effects of the chemical treatment were observed on abundance as well as on biomass. Effects were most pronounced 16 weeks after application of the test chemical (Figure 2.6). The observed effects on earthworm abundance and biomass did not differ between the TME tests and the respective field validation studies. Effects on earthworm diversity were difficult to assess since the number of individuals per species was usually too low. However, the genus *Lumbricus* and in particular *L. terrestris* and *L. rubellus* seemed to be more affected by the chemical treatment than others. EC50 values derived from the TME pretest, the TME ring test, and the field validation study indicate that the TME of the different partners delivered comparable results, although different soils were used. These results indicate that the abundance and biomass of earthworms are suitable endpoints in ecotoxicological studies with TMEs.

In the same project, Collembola were also studied, but only at 2 European sites (Koolhaas et al. 2004). Response of springtail communities was rather scattered, and no effects of carbendazim on species diversity were seen. Principal response curve

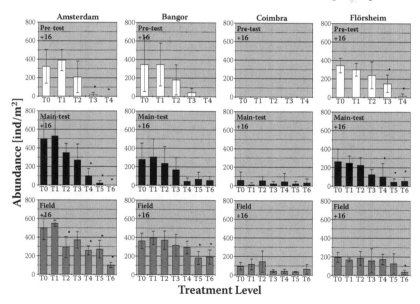

FIGURE 2.6 Effect of carbendazim on the abundance of earthworms (ind/m²). Data are given for the TME pretest, the TME ring test, and the field validation study; sampling point + 16 weeks after application; performed in Amsterdam, Bangor, Coimbra, and Flörsheim. Significant differences compared to the control are indicated by an asterisk. (from Römbke et al., *Ecotoxicology*, 13, 105–118, 2004.)

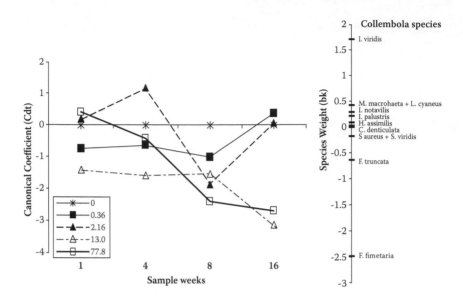

FIGURE 2.7   PRC diagram showing the effect of carbendazim on the Collembola community in a TME test 1, 4, 8, and 16 weeks after application. Concentrations tested were 0, 0.36, 2.16, 13.0, and 77.8 kg a.i./ha. (from Koolhaas et al., *Ecotoxicology*, 13, 75–88, 2004.)

(PRC) analysis demonstrated some significant effects of carbendazim on Collembola communities in 1 TME test and the field test (but not in the other TME test). In the Amsterdam TME test, *Isotoma viridis* appeared to be most sensitive to carbendazim treatment, while *Folsomia fimetaria* increased in numbers (Figure 2.7). Other Collembola species were hardly affected. In the field test, *Mesophorura macrochaeta* and *Friesea truncata* showed increasing numbers with carbendazim treatment; *Isotoma viridis* was again among the most sensitive species. NOECs for the effect of carbendazim on Collembola communities derived from these PRC analyses ranged between 2.16 and ≥87.5 kg a.i./ha. These results must be seen in the light of the expectation (based on laboratory tests) that this fungicide would not cause any effects on arthropods like Collembola. Further information about the use of the PRC approach to interpret the results of studies on the impact of pesticides on collembolans is given by Frampton et al. (2000) and Frampton (2001, 2007).

3 Legislative and Regulatory Background to the Assessment of Risks from Plant Protection Products in Soil

3.1 CURRENT REGULATORY POSITION IN EUROPE REGARDING SOIL TESTING FOR PPPS

At the time of writing, the regulation of plant protection products (PPPs) in the European Union (EU) is undertaken according to European Commission Directive 91/414/EEC and its various annexes and amending directives (European Council Directive 91/414/EEC 1991). This Directive is due to be replaced by the European Parliament and Council Regulation (EC) No. 1107/2009 (EC 2009). Although increasingly harmonized at a national level within Europe, certain European member states (MSs) have also required some further assessment of risks to soil organisms and functions when PPPs are registered locally. Other legislation will apply outside of Europe and to the assessment of risks from other forms of chemical contamination (e.g., contaminated land). While semi-field methods might well prove useful for such assessments, we concentrate here only on their potential for use in effects and risk assessment for PPPs within Europe.

EC Directive 91/414/EEC and its replacement Regulation 1107/2009 (EC 2009) concerns the placing of PPPs on the market, and they set out the ecotoxicological data requirements for both active substances and formulated products, as well as the circumstances in which these data are required. The ecotoxicological data requirements generally cover both acute and chronic effects and follow a tiered testing and risk assessment framework; i.e., they start with relatively simple acute laboratory tests and move on to more complex chronic laboratory, semi-field, and field tests, particularly for persistent or repeatedly applied substances. The data requirements for active substances related to soil testing are set out in Annex II to the directive, specifically annex points 8.4 (earthworms) and 8.5 (soil non-target microorganisms).

23

The equivalent data requirements for formulations (i.e., PPPs) are set out in Annex III to the directive, specifically annex points 10.6 (earthworms and other soil non-target macroorganisms) and 10.7 (soil non-target microorganisms).

Other organism groups that might inhabit soil are also covered in the annexes, in particular "arthropods other than bees" (Annex II, point 8.3.2; Annex III, point 10.5), and there exists the possibility to also consider "other non-target organisms (flora and fauna) believed to be at risk" (Annex II, point 8.6; Annex III, point 10.8). These last annex points are commonly used to consider the risks to non-target terrestrial plants. Again, while semi-field methods might well prove useful is assessing risks to some surface active arthropods and non-target plants, the focus here will be on those organisms that predominantly inhabit the soil profile.

The soil organisms and soil functions or processes actually mentioned in Annexes II and III to the existing directive are therefore earthworms, soil non-target microorganisms, other soil non-target macroorganisms (that contribute to the breakdown of dead plant and animal organic matter), and by inference, the impact on soil organic matter breakdown itself (Annex III, point 10.6.2). Among these, only the risk to earthworms is determined by directly considering effects on the organisms themselves, i.e., by testing and assessing "structural" effects on a single species in the laboratory (including lethal and sub-lethal effects), or on population and community structure and diversity under field conditions. Effects on soil microorganisms are determined largely through tests only on the processes of microbially mediated carbon and nitrogen mineralization in the laboratory, although the possibility of field testing remains. The impact on soil organic matter breakdown (and the organisms that contribute to it) is ultimately determined through testing the impact on the process itself under field conditions (e.g., the litter bag test), although in cases of intermediate soil persistence (DT90 field > 100 < 365 days) field tests can be triggered by lower-tier tests on individual soil taxa (earthworms, collembola, soil mites; see Figure 3.1). Such tests on soil processes themselves are often termed "functional" as opposed to "structural" (see Section 2.2). Research has suggested that no clear link can currently be established between structural effects and likely impacts on soil functions, so this link is intuitive rather than proven (Frampton et al. 2002).

It should be noted that ecotoxicological testing of soil-bound residues might also be triggered via the environmental fate and behavior annexes of EC Directive 91/414/EEC (Annex VI, point 2.5.1.1). Testing is always required where mineralization of a compound is <5% in conjunction with bound residue formation of >70% over 100 days. Guidance suggests that compounds with a high proportion of soil-bound residues should initially be treated as persistent (i.e., in this context, those with DT90 values > 365 days). Therefore, chronic earthworm tests and field tests on organic matter breakdown should be required. It is also worth noting that ecotoxicological testing is not restricted to the active substance and PPPs, but that tests on soil degradation and transformation products (commonly termed soil "metabolites") might also be necessary. This is often only required where metabolites are present at >10% of the initial dose of parent substance applied to soil in route or rate of degradation studies. In such circumstances testing in line with that on the active substance may be required unless other approaches can be adopted (e.g., arguments based on

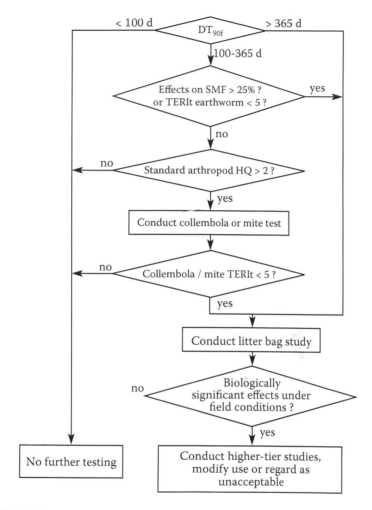

FIGURE 3.1 Test sequence to determine whether a litter bag test is required for persistent substances. SMF, soil microbial function. (From European Commission, *Guidance Document on Terrestrial Ecotoxicology*, Draft Working Document Sanco/10329/2002, rev. 2, final. October 17, 2002.)

structural activity relationships or the presence of metabolites at relevant levels in other ecotoxicity tests).

The "uniform principles" for evaluation and decision making when authorizing PPPs are set out in Annex VI of the directive (established under European Council Directive 97/57/EEC).

Specific evaluation and decision-making criteria for soil organisms are, however, only mentioned in Annex VI for earthworms and soil microorganisms, and these are unclear as to what precisely constitutes an unacceptable risk under field conditions. Acute and chronic toxicity exposure ratio (TER) triggers of 10 and 5, respectively, are mentioned for earthworms, and a 25% effect over 100 days trigger is mentioned for soil microbial functions. These are usually based on first-tier laboratory effects data and exposure predictions. Risk assessment criteria for other soil organisms and functions are currently not strictly defined in legislation. However, proposals have been made in other nonregulatory risk assessment schemes, such as those published by EPPO (2003).

So-called "unless clauses" in Annex VI allow for higher-tier determination of whether populations might be at risk following proposed use of the PPP under field conditions, although, as stated, the acceptability of any risk is not itself defined. Due to this lack of clarity in the directive and its annexes, help with interpreting the testing requirements and conducting the subsequent risk assessments is provided in a number of formal and informal guidance documents. Principle among these is the "Guidance Document on Terrestrial Ecotoxicology" under Council Directive 91/414/ EEC (SANCO/10329/2002) (EC 2002). Assessing the effects on soil organic matter breakdown is also discussed in detail in the SETAC "Guidance Document on Effects of Plant Protection Products on Functional Endpoints in Soil (EPFES)" (Römbke et al. 2003), where a number of the issues surrounding structural and functional soil testing have previously been aired (particularly in Chapters 2 and 9). Further "guidance" is also developed progressively during the scientific peer review of individual pesticidal active substances under the auspices of the European Commission and the European Food Safety Authority (EFSA). The EFSA scientific Panel on Plant Protection Products and Their Residues (PPR) has also issued a number of "opinions" that might constitute or contain guidance in the area of soil testing.

The "Guidance Document on Terrestrial Ecotoxicology" and other such papers have previously proposed that semi-field methods, such as TMEs, might constitute a potential higher-tier risk refinement step. In principle, these offer increased realism under more natural test conditions and allow for normal fate processes (e.g., dissipation of the test compound) and recovery of the soil organisms to occur. However, they also have increased complexity, duration, and costs, and potentially reduced reproducibility and precision compared with laboratory tests. Alternative refinement steps, such as additional single-species testing allowing for development of species sensitivity distributions (SSDs) or refined standard tests with more realistic exposure, might also be considered. It is, however, not clear precisely how any such methods would fit into the current tiered testing and risk assessment scheme under Directive 91/414/EEC or its replacement Regulation 1107/2009 (EC 2009).

Recently there has been increased activity and discussion regarding the need for semi-field test methods. For some there is a lack of a perceived need for such tests when existing methods of risk determination (including full-scale field tests) are thought to be sufficiently reliable, predictive, and cost-effective. Other discussion points include the lack of a suitable internationally agreed, validated, and standardized test method, and also the lack of clear soil protection goals and testing criteria within the existing directive. In particular, questions regarding whether the key protection goals for soil are structural or functional (or both) have regularly arisen during discussions on the future policy direction of regulation for PPPs. These same questions regularly arise in workshops such as EPFES and now PERAS. Until it is clearer what the regulatory testing procedure is intended to protect, under what circumstances, and what level of impacts and effects might be considered acceptable, it is difficult to determine appropriate test methods that might predict these effects with sufficient sensitivity and reliability while also remaining practical and cost-effective.

Many participants at PERAS considered that the current regulatory position, while not ideal in some respects, has evolved into a reasonably well understood testing and risk assessment strategy for PPPs. There was some understandable resistance to modify this position without good reason, and certainly while there was no significant movement or change in the policy-driven protection goals for soil. PERAS itself was not seen to be an appropriate forum to determine this policy direction. As discussed next, however, there may now be movement for change in the regulatory position for PPPs within Europe coming from a number of sectors.

3.2 POSSIBLE FUTURE REGULATORY DEVELOPMENTS IN EUROPE REGARDING SOIL TESTING FOR PPPS

The PERAS workshop took place at a time of change in the regulatory requirements for pesticides in Europe. Various proposals have been made (and views expressed) in relation to testing and risk assessment for soils. Although still under discussion, these are of clear relevance to the PERAS workshop and serve to underline why there is increased interest in higher-tier soil testing. The following section is based on a presentation given at PERAS and explains some of the reasoning behind the workshop and the challenges it faced. As this section points toward future developments, the situation discussed is bound to be provisional.

A number of developments are taking place that might determine the future direction of regulation and soil testing for pesticides. Most directly relevant is new Regulation 1107/2009 (EC 2009), replacing European Directive 91/414/EEC, and this will itself be heavily influenced by other relevant pieces of legislation, conventions, strategies, and decisions. Legislative influences include the Water Framework Directive (Directive 2000/60/EC 2000), the Biocidal Products Directive (Directive 98/8/EC 1998), the Dangerous Substances Directive (Council Directive 67/548/EEC 1967), and more recently, the Registration, Evaluation, Authorisation, and Restriction of Chemical Substances (REACH) Regulation (Regulation (EC) 1907/2006 2006). It is notable that some of these suggest a move toward the use of simple hazard cutoff criteria relating to persistence, bioaccumulation, and toxicity (PBT), and this might influence the authorization of PPPs, irrespective of the outcome of any use-based risk assessment. Closer to the topic of soil, the proposed Thematic Strategy for Soil Protection might also eventually lead to a European soil framework directive, although political progress on this is currently stalled. While use-based risk assessment will inevitably still feature in any future pesticides regulation, it is clearly the intention of both the European Commission and Parliament that it should encompass and harmonize with relevant aspects of other "upstream" legislation.

During PERAS (and the previous EPFES workshop) there was one aspect of PPP regulation for soils under Directive 91/414/EEC (and forthcoming Regulation 1107/2009) that led to intensive discussion. This was whether the protection goals for soil should be predominantly structural or functional, or both. While functional endpoints might allow protection of key processes in soil that enable it to retain its utility as an agricultural resource, they provide little information on the impact on individual taxa, populations, or communities of soil organisms. In part, this is due to high diversity and functional redundancy in soil; i.e., it might only take 1 or a few groups of organisms to maintain a functional process, and if any of these were adversely affected, they could still be replaced by others so that the overall function is not diminished (see Section 2.2 for further discussion). Differences of opinion were apparent among delegates at PERAS, some of whom considered that within the field environment, which is already severely impacted by operations such as plowing, the functional integrity of the soil was the ultimate protection goal. Off-field, it was recognized that different criteria might apply. Others felt that the in-field soil community was still part of the whole agricultural landscape and required protection of its structural diversity, particularly from longer-term effects and persistent compounds. Whether structural or functional, there was agreement at PERAS that the main protection goal (in-field at least) should be for recovery or recolonization from adverse effects to occur within 1 year (or cropping season) of initial treatment. Any semi-field method would, therefore, need to run sufficiently long enough to predict this recovery.

The aforementioned pieces of legislation give few clues as to how an EU soil protection strategy might develop in respect to protection goals. However, the main guidance document used in relation to the Water Framework and Biocidal Products Directives (as well as other EU new and existing chemicals legislation) is the European Commission's Technical Guidance Document (EC 2003). From this, Part II, Chapter 3, Section 3.6.2, "Strategy for Effects Assessment for Soil Organisms" states that "the objective of the assessment is to identify substances that present an immediate or delayed danger to the soil communities," and "the protection of the soil community requires protection of all organisms playing a leading role in establishing and maintaining the structure and the functioning of the ecosystem. The use of results from tests that represent different and significant ecological functions in the soil ecosystem is therefore suggested." This implies that both structural and functional effects are relevant protection goals. Another example of how the protection goals for persistent substances in soil might be addressed, at least at a national level, is the proposed Dutch decision tree for persistent pesticides (van der Linden et al., 2008b); see Section 3.3 for details.

The European Commission working group charged with revising and updating the data requirement annexes (currently II and III) of the Regulation to replace Directive 91/414/EEC has debated at length the information required for soil risk assessment. A view on this was also provided by the EFSA PPR in its "Opinion of the Scientific Panel on Plant Protection Products and Their Residues on a Request from the Commission Related to the Revision of Annexes II and III to Council Directive 91/414/EEC Concerning the Placing of Plant Protection Products on the Market—Ecotoxicological Studies" (adopted March 2007). In relation to soil testing, this opinion proposed the following key points, among others, which were clearly relevant to the PERAS workshop:

- The risk assessment for the terrestrial environment should be approached in a more integrated way, and include tests on structural endpoints (different species). Ecotoxicity testing (including field tests of, e.g., earthworms and non-target arthropods) should generally be carried out using a dose-response design rather than single-dose tests.
- The PPR Panel is of the opinion that, in particular, with respect to compounds that are persistent in soil, the terrestrial risk assessment should focus much more on the in-soil ecosystem structure than the current focus on the on-soil species. The in-soil assessment should make more use of testing with different species and taxa using structural endpoints, rather than continue to rely on the soil microorganism tests with only functional endpoints. The PPR Panel notes that there is a lack of experience with terrestrial field studies and semi-field studies, with respect to both the general understanding of relevant endpoints (structure, function, taxa, species, life cycle traits, diversity) and standardization and replication. Further research in these areas is needed, but this should not delay the formulation of test requirements.
- Given the relative scarcity of standardized test protocols for soil organisms, in contrast to the great diversity in survival strategies, a comprehensive assessment of the soil ecosystem is inadequate, even if all available test systems were employed. Test requirements should allow for the inclusion of nonstandardized test systems where standardized systems are unavailable.
- The litter bag test is a functional test and cannot protect the structure of the terrestrial community. The PPR Panel therefore supports the proposal not to include the litter bag in the data requirements, and to change the risk assessment approach accordingly.
- With respect to the omission of the previous data requirement on functional endpoints, the PPR Panel agrees that although the tests address ecologically relevant parameters, they have not proven to be of practical use in the assessments. Instead of discarding the assessment of soil microbial community, the PPR Panel suggests that effects on both functional and structural endpoints relating to bacteria, fungi, and protozoans, and also nematodes, should be considered. The use of other functional endpoints and other methods of assessing functional changes should also be considered.

It should be noted that, due to its founding remit, the EFSA PPR advises only on risk assessment and not necessarily risk management or political or societal protection goals. Its suggestions were therefore, couched within this limitation. The full PPR opinion is currently available via the EFSA Web site (http://www.efsa.europa.eu/) using the search term "EFSA-Q-2006-170."

At the time of the PERAS workshop, a draft of the revised ecotoxicology data requirement Annexes II and III had been made available in a public consultation by EFSA. Taking account of the PPR opinion and the views of EU member states and other stakeholders, the draft was then sent to the European Commission for further consideration and legislative progress. This draft proposed a more structural rather than functional focus for soil testing of PPPs, and there were clearer requirements for chronic testing. Most relevant to PERAS, the potential for use of higher-tier semi-field methods was explicitly mentioned in the revised draft Annex III. Still, at the time of writing, these revised data requirements annexes, along with new uniform principles (currently Annex VI) and the over-arching regulation itself, remain to be enshrined in EU legislation (i.e., the new Regulation (EC) 1107/2009).

At present, there might appear to be little scope for increased use of semi-field methods in soil risk assessment for PPPs. While there are occasions where further detailed consideration of effects on soil mesofauna (in particular) might be suitably addressed using a semi-field method, the existing suite of laboratory and field test methods appears to offer a practical solution to most of the questions that might arise.

If, however, the regulatory protection goals and data develop toward a greater focus on soil biodiversity and community structure for a wider range of soil meso-fauna (and possibly functions) than currently considered, then there is clear scope for the increased practical use of semi-field methods.

The semi-field test method itself needs to be further developed, and this is covered elsewhere in this publication. A key consideration from a regulatory perspective will be what regulatory questions can be answered by semi-field methods and what endpoints should be chosen to address these. It is likely that these might include a community no-observed-effect concentration (NOEC) or other threshold level, and experience should be gained from similar systems, e.g., aquatic mesocosms. There have been attempts to derive community NOECs for soil invertebrates before, but these have been limited by the lack of data in the appropriate format (Jänsch et al. 2006). How these refined endpoints should then be used in a subsequent risk assessment framework will also need further consideration, but this was beyond the scope of PERAS. For example, the question of whether an uncertainty or assessment factor is still required with end-points from such quasi-realistic studies arose repeatedly during the workshop. Ideally the triggers for moving from one tier of assessment to the next should also be validated against known acceptable or unacceptable effects seen in the field. It is likely, therefore, that additional guidance will be required in the future on how to design and make appropriate use of the output of semi-field methods in order to satisfactorily address regulatory questions. It is also likely, however, that the use of such methods will be relatively infrequent and be required to address specific questions related to the exposure and effects from compounds with particular properties and modes of action. Much use should be made of the output of lower-tier effect and environmental fate studies to help define the parameters that require further assessment. Therefore, PERAS agreed that any resulting guidance on test methods should not be too prescriptive and should allow for case-by-case protocols to be developed to answer the specific regula-tory questions that arise.

3.3 EXAMPLE OF THE REGULATORY USE OF HIGHER-TIER METHODS IN THE RISK ASSESSMENT FRAMEWORK FOR SOIL: THE DUTCH PROPOSAL FOR RISK ASSESSMENT OF PERSISTENT PLANT PROTECTION PRODUCTS IN SOIL

A summary of the Dutch proposal for risk assessment of persistent plant protection products in soil was presented at the PERAS workshop. A summary is also presented in these proceedings, since it provides an example only of how higher-tier methods might be used in a regulatory framework. Throughout the workshop several elements of the Dutch proposal were discussed as examples of how certain aspects could be handled, such as the protection goals and the classification of the effects found in (semi)field studies.

This proposal, along with other options, is likely to be discussed during revision of 2 key guidance documents relating to higher-tier risk assessment for pesticides in soil. These are the "Guidance Document on Terrestrial Ecotoxicology" (SANCO/10329/2002) and the "Guidance Document on Persistence in Soil" (European Commission, 9188/VI/97, rev. 8, December 7, 2000). At the time of writing (spring 2009), EFSA working groups have been established to develop these documents in light of recent regulatory changes.

3.3.1 INTRODUCTION

Persistence in soil is one of the evaluation aspects of plant protection products. However, except for trigger values indicating persistence in soil, there is no broadly accepted evaluation procedure at the European level, and member states use different approaches for the evaluation of persistence in soil at the national level. In the Netherlands a methodology for risk assessment of plant protection products for persistence was proposed (Van der Linden et al. 2008a, 2008b; see http://www.rivm.nl/bibliotheek/rapporten/601712003.html and http://www.rivm.nl/bibliotheek/rapporten/601712002.html). The approach has been developed for the in-crop area.

The proposal considers 3 protection goals (Brock et al. 2006):

Protection of life support functions of the in-crop soil to allow the growth of the crop and protection of key(stone) species (earthworms) of agricultural soils (in line with the so-called "functional redundancy principle" (FRP)). This protection goal is already assessed at the European level according to existing requirements (European Council Directive 91/414/EEC 1991, due to be replaced by Regulation 1107/2009; EC 2002). Therefore, it is not discussed further here.

Protection of life support functions of the soil to allow crop rotation and sustainable agriculture, with overall protection of the structure and functioning of soil communities characteristic for agroecosystems (in line with the so-called "community recovery principle" [CRP]).

TABLE 3.1

Proposed principles to set protection goals for in-crop soils, trigger values, and time window

Principle to set protection goal	Time window	Trigger		
		DT50 > 30 d	DT50 > 90 d	DT50 > 180 d
Functional redundancy principle (FRP)	In year of cropping	Testing according to FRP	Testing according to FRP AND	Testing according to FRP
Community recovery principle (CRP)	Two years post last application		Testing according to CRP AND	Testing according to CRP
Ecological threshold principle (ETP)	Seven years post last application			Testing according to ETP

Protection of life support functions of the soil to allow changes in land use, with overall protection of the structure and functioning of soil communities characteristic for nature reserves (in line with the so-called "ecological threshold principle" (ETP)).

The approach has been developed for the in-crop area. The protection goals take effects at different moments in time post last application:

FRP in year of cropping
CRP 2 years post last application
ETP 7 years post last application

Table 3.1 presents a scheme that shows the relation between the flowcharts and the principles to set protection goals as triggered by different DT50 values. Since the protection goals in line with the different principles are used for different points in time post last application, there is no a priori hierarchy for the different goals. It is not deemed necessary to test all protection goals for all compounds, since different triggers (DT50 values) for the different protection goals are proposed.

The procedure provides trigger values for the half-life for dissipation (DT50) from soil (see Table 3.1). Separate decision schemes were proposed for both protection goals. In these schemes both the predicted environmental concentrations (PECs) and the ecotoxicological endpoints can be determined using tiered approaches.

3.3.2 Tiered Approach for the Exposure Assessment

Exposure concentrations in test systems are essential for deriving ecotoxicity endpoints. Only rarely is all essential information on environmental conditions and substance properties available for these test systems. Therefore, in the proposal procedures are described to derive conservative estimates for the exposure concentration. Apart from the exposure in the test system, a tiered assessment is proposed for exposure in the risk assessment. In the first tier, simple models and realistic worst-case assumptions are used, resulting in a realistic worst-case PEC. In the higher-tier assessment, the GeoPEARL model (www.pearl.pesticidemodels.eu/) is used with more realistic assumptions, resulting in a refined PEC_{soil}.

The results of the tier 1 calculations, i.e., the total available content in soil or the pore water concentration as calculated for the realistic worst case with the simple model, are compared with the results of one of the effect tiers (see next section). If the exposure resulting from tier 1, compared to the effect tiers, results in unacceptable effects, the assessor may decide to go to the second tier.

The result of the second tier is the realistic worst-case exposure for the area of use of the plant protection product. The realistic worst-case exposure here is defined as the spatial 90th percentile of the available total contents in soil or the available pore water concentrations after 20 periodic application regimes.

3.3.3 TIERED APPROACH FOR THE EFFECT ASSESSMENT

As a first-tier approach for the CRP, it is proposed to base the permissible concentration on the long-term toxicity exposure ratio (TER) on a basic set of standard soil organisms with a long-term TER > 10. For the ETP a TER of 100 is proposed in the first tier.

As a second tier, the species sensitivity distribution (SSD) approach is proposed as well as the calculation of the median hazardous concentration for 5% of the species (HC_5) (Aldenberg and Jaworska 2000; Posthuma et al. 2002; Frampton et al. 2006). For the ETP it is proposed to compare the PEC with the lower limit of the 95% confidence interval of the HC_5.

As a third tier, the performance of semi-field tests is proposed. "Effect classes" (adapted from De Jong et al. 2005; Brock et al. 2006) could be used to facilitate the interpretation of concentration-response relationships for relevant measurement endpoints of terrestrial semi-field experiments, namely,

Class I: No treatment-related effects
Class II: Slight treatment-related transient effects, usually on 1 or a few isolated sampling dates only
Class III: Clear effects on several consecutive sampling dates, lasting less than 2 months post last application of the PPP in the test system
Class IV: Clear effects on several consecutive sampling dates, lasting longer than 2 months, but full recovery within a year post last application of the PPP in the test system
Class V: Clear long-term effects; full recovery not within 1 year post last application of the PPP in the test system

Since the protection goals are used for different points in time post last application, there is no a priori hierarchy for the different goals (see Table 3.1). It is proposed to consider an exposure concentration at 2 and 7 years post last application, acceptable if this exposure concentration results maximally in class I or II effect responses in an appropriate semi-field test. Consequently, the main focus is on threshold concentrations for effects derived from (semi)field tests, and these values are used to address the occurrence of potential recovery after 2 years. An extra assessment factor (AF) may be applied to overcome the remaining uncertainty with respect to spatial extrapolation of the effect assessment based on a single semi-field test. To date, too few terrestrial semi-field experiments with the same PPP have been performed to scientifically underpin the magnitude of such an extra AF. Based on the calculated uncertainty in the geographical extrapolation of threshold levels for effects observed in aquatic micro- and mesocosms with PPPs (cf. Brock et al. 2006), however, an appropriate AF might be 3. For the ETP an (arbitrary) extra factor of 3 is applied in order to cope with the differences between agroecosystems and ecosystems under more natural conditions. Further research and experience with terrestrial higher-tier studies is needed to specify the AF. One of the recommendations is the development of (semi)field methods. In one of the case studies of the Dutch proposal, using carbendazim, TMEs were used to derive a higher-tier endpoint.

3.3.4 Semi-Field Tests with Carbendazim

As one of the case studies, the proposed method was tested with the available data for fungicide carbendazim (for details see Van der Linden et al. 2008a). Several ecotoxicological (semi)field tests are reported for carbendazim in a special issue of the scientific journal *Ecotoxicology* (see Knacker et al. 2004). The reported (semi)field tests comprise studies using indoor terrestrial model ecosystems (TMEs) and corresponding outdoor field plots representative of 4 different European sites: Amsterdam (the Netherlands), Bangor (United Kingdom), Coimbra (Portugal), and Flörsheim (Germany). These (semi)field experiments were considered appropriate for use in the risk assessment procedure on the basis of the following criteria:

1) The test systems represented a relevant soil community.
2) The setup of the experiments was adequately described.
3) The exposure regime in the test systems was well characterized (although a detailed evaluation needs the basic data that underlie the scientific publications).
4) The investigated species, particularly Enchytraeidae and Lumbricidae, are reported to be sensitive to the fungicide carbendazim (although structural aspects of soil fungi were not investigated).
5) It was possible to evaluate the observed effects statistically and ecologically (univariate and multivariate techniques).

The structural measurement endpoints investigated in the TME mainly concerned soil invertebrates. Treatment-related effects on soil microorganisms were only investigated from a functional point of view (microbial activity like nutrient cycling and carbon mineralization).

TABLE 3.2

Lowest and geometric mean calculated exposure concentrations in the upper 5 cm soil layer of the 4 TMEs

Type of exposure concentration	At time (d)	Terrestrial model ecosystems			
		Lowest value		Geometric mean value	
		Effect Class I NOEC	Effect Class II LOEC	Effect Class I NOEC	Effect Class II LOEC
PEC_{soil} total content (mg kg^{-1})	0	0.54	0.72	0.86	1.26
PEC_{soil} total content (mg kg^{-1})	42	0.22	0.46	0.45	0.73
PEC_{TWA42} total content (mg kg^{-1})		<0.58	0.58	0.63	0.98

Note: NOEC and LOEC values are based on the most sensitive measurement endpoint; in this specific case, the community of Acari, PEC_{TWA42} = time-weighted average PEC over 42 days, based on the duration of the test with the most sensitive organism.

For a proper effect and risk assessment the (semi)field threshold levels in kg ha^{-1} were recalculated to obtain higher-tier NOEC or LOEC values for the soil invertebrate community in terms of concentration in the upper 5 cm of soil. From the results it can be concluded that in the indoor terrestrial model ecosystem arthropods (Collembola, Acari) and earthworms and potworms were among the most sensitive measurement endpoints.

Table 3.2 presents a summary of the effect class I and II threshold concentrations in the TMEs that were used in the higher-tier risk assessment. In this table the lowest values reported for the 4 TME are given, as well as the geometric mean values for all the TMEs. In calculating these geometric means, "larger than" and "smaller than" values were not used. In the TME the difference between effect class I NOECs and effect class II LOECs is relatively small.

A bottleneck in the risk assessment of carbendazim according to the proposed procedure is that, although the substance is a fungicide, hardly any information is available on the impact of carbendazim on densities of soil fungi and/or the composition of the fungal community in soils. This is the case for the first-tier as well as for the higher-tier assessment.

4 Overview and Evaluation of Soil Semi-Field (Higher-Tier) Methods

4.1 SCOPE

The use of semi-field (higher-tier) methods aims to implement ecological realism into risk assessment methodologies (for a definition, see Figure 1.1 and Section 4.2). First, Odum (1984) described mesocosms as "bounded systems, partly permeable to their surroundings" and gave therewith the most basic definition of a semi-field approach. He proposed them for use in both ecological and ecotoxicological research. Historically, they are based on approaches developed for ecological questions (e.g., Verhoef 1996; Fraser and Keddy 1997), but as early as the in the mid-1980s a terrestrial model ecosystem was proposed as an ecotoxicological test (Van Voris et al. 1985; Sheppard 1997). These systems can provide improved effect data to evaluate single-species test results and can help to measure ecosystem functions under controlled conditions. They can be used to determine indirect as well as synergistic or compensatory effects of chemicals at ecosystem level and allow for significantly improved assessment of the fate of contaminants in terrestrial ecosystems. Semi-field (higher-tier) methods are designed in a way that the advantages of laboratory tests (e.g., standardization, controlled conditions) are combined with the advantages of field studies (natural variability, complex interactions), while at the same time avoiding their disadvantages, like focus on single species or high amount of man power, respectively. In short, while these tests focus on the biological organization level of the population and community, they cover a very wide range of methodological approaches (Figure 4.1).

In order to facilitate communication, we can separate soil ecotoxicological studies into 3 experimental levels, defined as follows:

Laboratory tests: Experiments in which the impact of a substance is studied under controlled conditions (both concerning exposure and environmental variables). Usually, they are characterized by "unrealistic" conditions (e.g., spiking of the substance into an artificial test substrate), short durations (some days up to 2 months), and a focus on few standard species, which have been selected mainly for reasons of practicability.

Experimental field studies: In agreement with Liess et al. (2005), they are experiments analyzing the impact of a substance applied under controlled

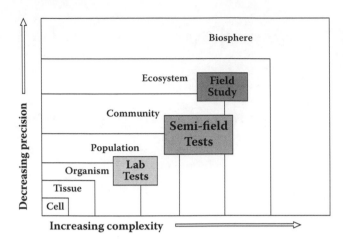

FIGURE 4.1 Place of multispecies semi-field tests in relation to different levels of biological organization.

conditions. Such studies are performed in the natural environment within an agricultural context, often focusing on 1 organism group, in particular earthworms. It should be noted that, in this case, "controlled conditions" refers to the application of the substance but not to the environmental variables.

Semi-field tests: They are defined as controlled, reproducible systems that attempt to simulate the processes of and interactions between components in a portion of the terrestrial environment, either in the laboratory (small scale) or in the field, or somewhere in between. In Chapter 5, this description will be outlined in detail, using some examples.

Thus, this chapter will focus on an overview of which methods are available, how they can be classified, and how they can be evaluated according to criteria that basically do not differ from criteria for other ecotoxicological methods (see Section 4.4 for details).

4.2 CLASSIFICATION OF EXISTING SEMI-FIELD APPROACHES

As mentioned earlier, the term "semi-field test" covers all methods between the field and the laboratory levels. In order to group semi-field tests, the following dichotomic criteria have been used, taking proposals from Morgan and Knacker (1994) and Römbke and Moltmann (1996) into consideration:

a) Soil integrity
   Has the natural soil been modified or not?
   Intact soil cores vs. modified or assembled soil (by sieving and/or defaunating)
b) Source of organisms

Have the native organisms been used or have individuals of selected species (e.g., standard test species) been added? A mixed approach is imaginable where natural fauna is "enriched" or supplemented by reared species (in order to simulate trophic levels or add traits still lacking).
Natural community vs. added individuals

c) Climatic conditions
Is the study performed under controlled environmental conditions or not?
Field situation vs. controlled scenario

d) System integrity/degree of isolation
Is the test system connected with its environment, i.e., (i) is an exchange of the test substance possible or not (actually, mainly important for radio-labeled substances), and (ii) is an exchange of biota possible, i.e., is the recovery of community endpoints strictly intrinsic or not? Systems can be closed and gas-proofed by plastic lids, or immigration of larger animals is prevented by gauze of different mesh sizes.
Open vs. closed systems

e) Size
How large is the test system? Since this criterion could not be used in a dichotomous way, it has not been used for the final classification.

Despite the fact that no classification approach is able to cover all potential semi-field test systems, the following typology is proposed. It refers mainly to criteria a, b, and c.

A	Assembled soil systems	Artificially assembled units with added organisms (alone or in combination with remnants of the natural community, such as nematodes)
	A1	Controlled environmental conditions
	A2	Field conditions
B	Terrestrial model ecosystems (TMEs)	Intact soil cores with natural communities
	B1	Controlled environmental conditions
	B2	Field conditions
C	Field enclosures	Undisturbed soil, immigration of species prevented by barriers
	C1	Natural communities
	C2	Added organisms

It should be kept in mind that not all potentially possible combinations of these 3 criteria could be covered by examples in the following, since only very few studies belonging to A2 or C1 have been performed so far. In addition, "grey zones" exist between the investigation levels:

Grey zones between laboratory and semi-field level:
• Laboratory tests with 2 species, e.g., predatory mites and collembolans (Axelsen et al. 1997)

- Microcosms focusing on fate endpoints, e.g., influence of plants on the degradation of a pesticide (Schuphan et al. 1987)

Grey zones between semi-field level and field level:

- Combination of laboratory and field phases in 1 test, e.g., with staphylinid beetles (Metge and Heimbach 1998)
- Large lysimeters focusing on leaching behavior under (almost) field conditions (Führ and Hance 1992)

4.3 PRESENTATION OF SEMI-FIELD METHODS

The literature review performed focused on ecotoxicological methodological papers. In total, approximately 150 papers were identified, including a high number of "grey" reports, which—in part—still need to be assessed.

Out of those, 51 papers on semi-field methods were evaluated in detail. From these, 34 papers focused on assembled soil systems (group A) (70%), 10 papers on terrestrial model ecosystems (group B) (20%), and only 7 papers on field enclosures (group C) (10%). Four examples (ISM (Burrows and Edwards 2002), TMEs indoors (Knacker et al. 2004) and outdoors, carabid beetle test (Heimbach et al. 2000)), representing the 3 main groups, are shown. In addition, more references are given for all 6 groups. The features of the different semi-field approaches in relation to simple single-species tests and full-scale field studies are summarized in Table 4.1.

Group A1: Artificially assembled systems under laboratory conditions

These are usually relatively small systems, often also called gnotobiotic tests (Morgan and Knacker 1994; Scott-Fordsmand et al. 2008). It is assumed that all organism groups of the system are under control and are well known by the experimenter (*gnostos* (Greek) = known). That does not imply that, in some experiments, these groups (e.g., nematodes) are inevitably unknown. In these cases they have to be analyzed

TABLE 4.1
Categorizing criteria for terrestrial ecotoxicological test methods

Criterion	Assembled systems	Terrestrial model ecosystems	Field enclosures
Soil integrity	Sieved	Intact	Intact
Composition of test species	Typical food chain	Natural community	Natural community
Origin of test species	Lab culture	Site of origin	Site of origin and lab culture
System integrity	Open or closed	Open or closed	Open
Environmental control	Indoors or outdoors	Indoors or outdoors	Outdoors
Position in tiered approach	Higher tier	Higher tier	Higher tier

by measurements or by taxonomic identification after the application of a chemical.

This category is especially suited to follow fate and behavior of chemicals through soil, air, aquatic, and biotic compartments of an ecosystem (so-called "vegetation chambers" by Schuphan (1986)). Radiolabeled compounds were often applied and attempts were made to simulate artificial food chains.

Research in this area started in the early 1970s (Metcalf et al. 1971; Cole et al. 1976) and was continued mainly by the work with "terrestrial microcosm chambers" of Gile (see, for example, Gillett and Gile 1976; Gile et al. 1980; Gile 1983; and other publications of this group).

Other early attempts to use artificial systems were done, e.g., by Van Wensem et al. (1991). They called their systems microecosystems (MESs) and measured functional endpoints such as poplar leaf litter decomposition and mediated microbial activity by the isopod *Porcellio scaber.*

An example of this group of systems is the integrated soil microcosm (ISM) test, which is also known as the Ohio approach (Edwards et al. 1996). Actually, it is not absolutely typical since, while most of the test organisms are added, the microbial community and nematodes are part of the original community living in the sieved soil. Its main features are summarized in Table 4.2 and in Figure 4.2.

Note: In order to improve the moisture regime as well as the collection of leachate, tension could be applied at the bottom of each soil column (30 to 35 kPa) to mimic field conditions (Checkai et al. 1993), which can also be used in other semi-field tests systems in both groups A and B (Figure 4.3).

TABLE 4.2

Main features of the ISM test as described by Edwards et al. (1996) as an example for semi-field group A1

Name	ISM test (integrated soil microcosm)
Guideline/literature	Burrows and Edwards (2004)
Principle	Testing of natural and added organisms in sieved field soil; performance of leachate tests possible
Species	Natural soil microbial, nematode, and microarthropod (partly) community, added plants and invertebrates (mainly earthworms)
Substrate	Sieved field (mainly agricultural) soils
Duration	Usually 21–28 days
Parameter	Wide variety of fate and effect endpoints
Experience	So far mainly pesticides and explosives

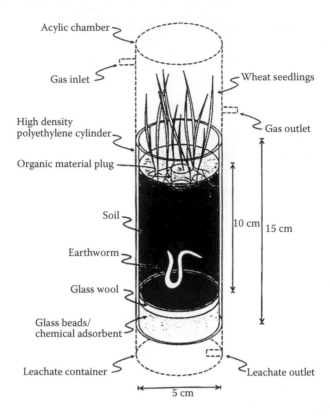

Acylic chamber

Gas inlet

Wheat seedlings

High density
polyethylene cylinder

Gas outlet

Organic material plug

Soil

10 cm | 15 cm

Earthworm

Glass wool

Glass beads/
chemical adsorbent

Leachate container

Leachate outlet

5 cm

FIGURE 4.2 Schematic view of the ISM test. (In Kuperman et al., in *Environmental Analysis of Contaminated Sites*, ed. Sunahara et al., John Wiley & Sons, Chichester (UK), 2002, p 45–60.)

Further examples of this type of semi-field study are the MS•3 test (Fernandez et al. 2004; Boleas et al. 2005), the soil multispecies test system (SMS) test (Scott-Fordsmand et al. 2008), or slightly larger systems called mesocosms (Pernin et al. 2006).

Group A2: Artificially assembled systems under field conditions

Since this kind of test system is rarely used (e.g., Løkke 1995; de Vaufleury et al. 2007), no detailed example will be presented here.

Group B1: Intact soil cores with natural communities (TMEs) under laboratory conditions

The most important example of this group is terrestrial model ecosystems, already developed more than 20 years ago (Van Voris et al. 1985), and still the only standardized terrestrial semi-field method (ASTM 1993; EPA 1996) (Table 4.3, Figures 4.4 and 4.5). The fungicide carbendazim and other pesticides have been intensively studied with TME in Europe (Förster et al. 2004; Knacker et al. 2004) and Brazil (Förster et al. 2006).

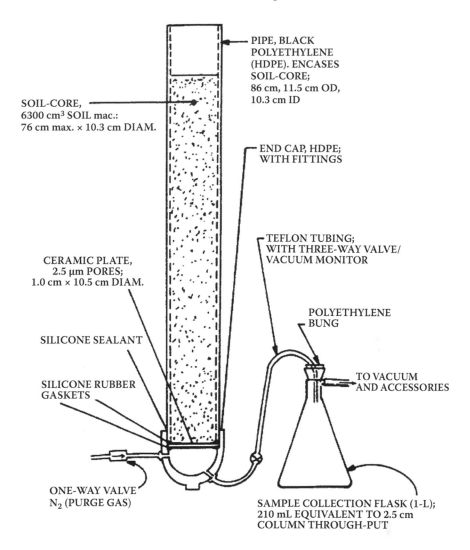

FIGURE 4.3   Soil column with water collection (Checkai et al. 1993).

Group B2: Intact soil cores with natural communities (TME) under field
conditions
     Similar soil cores have been used recently under field conditions that
are considerably larger (up to 47 cm in diameter, 40 cm height, 100 kg
of grassland soil) than those kept under laboratory conditions. They were
stored outdoors in a facility that allows for controlled moisture conditions
through irrigation or shielding (Table 4.4, Figures 4.6 and 4.7). It has been
shown that population dynamics of most dominant species follow natural
fluctuations over the years (Scholz-Starke et al. 2008).

TABLE 4.3

Main features of the TME test as an example for semi-field group B1

Name	Terrestrial model ecosystems (TMEs)
Guideline or literature	ASTM (1993), UBA (1994), USEPA (1996), Knacker et al. (2004)
Principle	Interaction of soil properties and the natural community of microorganisms, animals, plants
Species	Natural soil organism community
Substrate	Undisturbed soils from field sites
Duration	Usually about 16 weeks
Parameter	Wide variety of fate and effect endpoints
Experience	Growing experience: e.g., with fungicides, contaminated field soil, or pharmaceuticals in dung

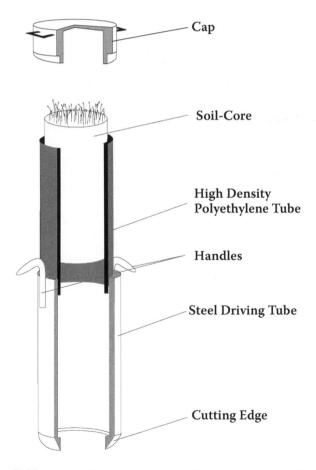

Cap

Soil-Core

High Density
Polyethylene Tube

Handles

Steel Driving Tube

Cutting Edge

FIGURE 4.4 TME cross section and extractor. (Figure designed by Thomas Knacker. [UBA 1994].)

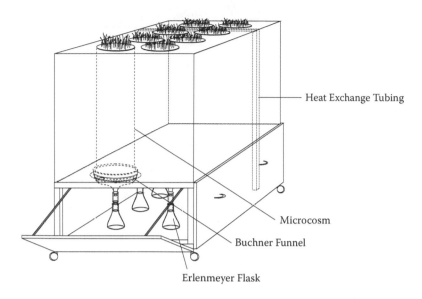

Heat Exchange Tubing

Microcosm

Buchner Funnel

Erlenmeyer Flask

FIGURE 4.5   TME in chart with leachate funnels. (Figure designed by Thomas Knacker.)

TABLE 4.4

Main features of the TME test as an example for semi-field group B2

Name	Terrestrial model ecosystems (TME)
Guideline/literature	Not yet available
Principle	Natural dynamics of original soil communities taken from undisturbed grassland
Species	Natural soil organism community
Substrate	Undisturbed soils from field sites
Duration	Proven to be stable up to 1 year
Parameter	Community structure of Collembola, Oribatida, Enchytraeidae, Nematoda, plant biomass; fate measures can be included
Experience	Growing experience mainly with insecticidal (and some fungicidal) compounds

The following 2 groups are summarized as "field enclosures." Usually a steel frame is rammed into the soil, so soil structure remains undisturbed. Homogenized soil has been very rarely used. A single example of using sieved, defaunated soil that has been reinoculated subsequently by adding nematodes is given by Smit et al. (2002). The effect of zinc contamination on nematode communities was investigated. In ecological research, there is much more experience with all possible combinations of building up field enclosures (open or isolated, sieved or undisturbed, natural communities or inoculated enclosures). Kampichler et al. (1995, 1999) measured different functional and structural endpoints in systems called "field mesocosms" and gained many insights of stability and ecology of enclosed systems.

FIGURE 4.6   The surface area of a TME (diameter 47 cm) allows for sequential coring of up to 19 subsamples for extracting enchytraeids or microarthropods and several small cores. (Photo by Bernhard Theißen.)

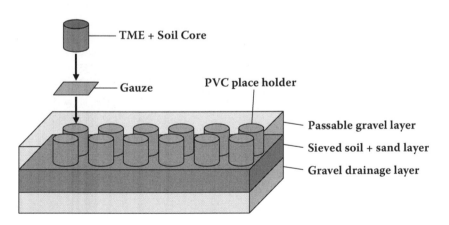

FIGURE 4.7   Outdoor experimental facility. (Figure designed by Andreas Toschki.)

TABLE 4.5

Main features of the Carabid semi-field test as an example for group C2

Name	Carabid semi-field test
Guideline/literature	Heimbach et al. (2000)
Principle	Testing of acute effects of pesticides on selected organisms under field conditions
Species	*Poecilus cupreus* (Carabidae; ground beetles) from laboratory cultures
Substrate	Enclosures at undisturbed field sites
Duration	Usually 14 days (check every 3 days)
Parameter	Mortality and feeding rate of beetles
Experience	Several pesticides; few publications

Group C1: Field enclosures with natural communities

Currently, there are only few examples known for this group of semi-field methods, but enclosures in wheat fields using field catches of collembolans (Wiles and Jepson 1992) may come closest. Frampton and Wratten (2000) compared effects of fungicides on collembolans in winter wheat in barrier-enclosed and unenclosed plots.

Group C2: Field enclosures with added organisms

While this group of methods has rarely been used with endogeic soil organisms (one exception: earthworms (Callahan et al. 1991)), it is a quite common approach when studying the effects of pesticides on non-target arthropods living on the soil surface (Metge and Heimbach 1998; Candolfi et al. 2000; Heimbach et al. 2000).

In Table 4.5 and Figures 4.8 to 4.10 the main properties of the carabid semi-field test are described.

4.4 EVALUATION OF SEMI-FIELD METHODS

In order to evaluate the suitability of the presented semi-field tests to be used as a higher-tier system in the context of the risk assessment of pesticides, 2 groups of criteria were used, covering aspects of ecology and performance (Römbke et al. 1996). The following compilation reflects the outcome of this discussion:

4.4.1 ECOLOGICAL CRITERIA

Relevance: The system should include important species (e.g., ecosystem engineers, keystone species, sensitive species), or in the case of multispecies tests systems, the species composition should represent the community of the habitat of concern, e.g., open land communities, that is typical for the agricultural landscape.

Endpoints: Total number, covering structure and function.

Flexibility: Suitable for different exposure scenarios, different soil types, and different crops.

Sensitivity: Sensitive to chemicals but robust toward other factors. The system should react in a relevant dose range. The system should contain sensitive species.

FIGURE 4.8   Carabid semi-field test: overview. (Photo by Jörg Römbke.)

FIGURE 4.9 Introduced species: Carabid *Poecilus cupreus*. (Photo by Andreas Haller.)

FIGURE 4.10 An individual enclosure of the carabid semi-field test. (Photo by Jörg Römbke.)

4.4.2 PERFORMANCE CRITERIA

Practicability: Good ratio between resources (costs, time, staff) and results.
Reproducibility and repeatability: Statistical robustness, i.e., low variability of
chosen endpoints.
Experience: Amount of studies performed including field comparisons.
Standardization: Guideline or guidance paper available.

Since it is almost impossible to quantify the degree of fulfillment of these criteria,
it was decided to use expert knowledge in order to perform this evaluation process.

Analyzing Table 4.6, it becomes obvious that there is no "best use" method for
all research or assessment questions concerning higher-tier testing of chemicals.
It has to be decided case by case which method is most appropriate. For instance,
if an impact on 1 important predator species is of interest, a field enclosure is a

---

TABLE 4.6
Rough classification of 3 groups of semi-field methods (assembled systems, terrestrial model ecosystems, field enclosures) according to 8 ecological and performance criteria as indicated by 3 shades of gray and additional comments

		Assembled systems	Terrestrial model ecosystems	Field enclosures
Ecological criteria	Relevance	Artificial food chain, but no real competitors or prey-predators	Natural community	Increased density of predators
	Endpoints	No community measures	All "known" parameters can be measured	All "known" parameters can be measured
	Flexibility	No crop simulation possible	Most soils, except very sandy or very dense soils	All soils
	Sensitivity			
	Practicability			
Performance criteria	Reproducibility and repeatability		Exact result not reproducible because of natural plasticity	Several studies, but few publications
	Experience	If similar approaches are combined	EU ringtest	
	Standardization	None	ASTM guideline and UBA draft available	IOBC guideline available

High/good/many	
Medium/fair/numerous	
Low/bad/few	

suitable choice. On the other hand, if mainly structural endpoints are of interest and indirect effects along the food web are expected, a TME study may be preferable.

4.5   RECOMMENDATIONS

According to the outcome of the review in the previous section, the following main statements are possible:

- Several different options for semi-field testing are available, which may offer a range of potential tools for higher-tier environmental risk assessment of pesticides in soil. So far, the most experience has been gained using TMEs that have been demonstrated to provide reliable and reproducible data concerning effects of pesticides on soil organisms.
- The selection of the most appropriate higher-tier method (laboratory, semi-field, or field) depends on the research or risk assessment requirements and the regulatory question that needs to be addressed.
- When designing a semi-field study, it is important to consider both the ecology of key species under investigation and the fate and behavior of the test substance.

5 Technical Recommendations

In the following, the discussions held in the 4 "technical" working groups of the PERAS workshop are summarized. Formally, this compilation is based on the reports presented by the 4 workgroup rapporteurs, as well as on the overall summary presented by the workshop rapporteur. The information provided here mirrors the agreements made in the 4 groups. In those cases where no agreement could be achieved, the differing opinions are presented. Methodological and detailed technical recommendations for the design and performance of TME studies are given in Appendix 2.

The discussions in this chapter focus almost exclusively on the TME method because there is more experience relating to its use, and more data related to its use with pesticides, than there is for other methods. However, the majority of the issues and recommendations discussed may also apply to higher-tier investigations using other semi-field methods.

5.1 FATE AND EXPOSURE

Research is necessary with respect to the application of persistent pesticides, as further outlined in Chapter 6. Readily degradable pesticides (DT90 < 100 days according to SANCO/10329; EC 2002) should be applied to the soil semi-field systems according to good agricultural practice (GAP) as close as possible in terms of concentration, application pattern, application technique, and seasonal considerations. Alternatively, a dose-response test strategy might be suitable to answer specific questions.

Based on current practice for persistent compounds (DT90 > 100 days), however, the method of pesticide application should take into account that a plateau concentration might be established in the top soil layer, due to previous applications of the pesticide, resulting in accumulation in soil. At tier I risk assessment for soil organisms and functions, the effects endpoint is typically compared with the peak PEC_{soil} plateau (at 5 cm depth), i.e., the baseline plateau concentration plus the concentration resulting from the total in-year dose. However, a tier II refinement may consider a baseline PEC_{soil} plateau based on greater soil depths (e.g., 20 cm), where this can be fully justified due to the method of application (i.e., soil incorporation) or the particular cropping system. This dilution of the active substance (or soil metabolite) PEC may, for example, be chosen because it is considered that during the accumulation period, the agricultural soil is likely to be deep plowed. Such an approach would not be appropriate where minimal or no-tillage practices might be followed. In Chapter 6, the different methods of applying persistent pesticides are described:

either the simultaneous addition of the accumulation plateau concentration in the top soil layer, e.g., 5 cm, and the annual dose, or a preliminary aging of the added plateau concentration in the soil and a later application of the annual dose according to good agricultural practice. The impact on soil organisms of directly incorporating these concentrations at different depths rather than relying on "natural" movement, irrigation, and mixing into the soil profile might also be considered.

Soils in semi-field tests should always be covered with plants. If the compound is normally applied to bare soil, then plants should be sown subsequently to reflect the GAP. Independent of the method used, and the type of pesticide tested, it has to be ensured that the test compound will reach the soil, e.g., by irrigation of the crop to wash off the plant-bound ingredient to the top soil layers. Likewise, it has to be ensured that the organisms in soil are exposed to the test substance. Irrigation of the soil should be adapted to regional circumstances and according to the cover crop and faunal demands to ensure the optimal moisture content. Soil temperature should be monitored for semi-field tests under field conditions, while in glasshouse or labora-tory facilities, a range of 15 to 20 °C should typically be established. Also, the light conditions should be measured in case of potential photolysis of the test compound on the soil surface and with respect to plant growth.

It was debated during the workshop whether persistent pesticides should be applied to agricultural (i.e., arable) soils only, while nonpersistent pesticides might be applied either to grassland or arable soils. However, studies addressing a risk assessment of in-crop concerns may need a different case-by-case approach than studies for off-crop concerns because of differences in soil cover.

To evaluate the suitability of the semi-field test systems with respect to the sensi-tivity of the soil biota, toxic (positive) controls should be applied to separate systems. It was also argued that no standard soil should be defined to be used in semi-field tests. However, recommendations for ranges of relevant soil characteristics should be established.

Analytical confirmation of the added pesticide concentrations should be achieved in the same soil strata that will be used for the ecotoxicological effect testing. Modeling the fate of the test substance according to data from other relevant tests might be an alternative instead of chemical analyses, but this needs to be related to likely exposure in the actual test system.

In order to enable extrapolation of the test results to other environmental condi-tions, on both a spatial and a temporal scale, different soils, soil moistures, seasons, etc., should be tested. Ideally, such generic research might establish the degree to which test results can be extrapolated to different climatic regions and soil types.

5.2 EFFECTS

Within this PERAS working group, several topics considering predominantly the ecological relevance of TMEs in mimicking field situations were discussed.

There was a consensus that TMEs might appropriately mirror field situations, given that potentially sensitive soil organism groups are present at sufficient abun-dances within the cores and that they are exposed to the test substance. It was rec-ognized that some organism groups, mainly epigeic macrofauna species (beetles, isopods, spiders, snails, lumbricids), may not be adequately represented in the TME, and that soil micro- and mesofauna (e.g., nematodes, microarthropods, enchytraeids) and soil microorganisms are the key groups targeted by this system. The issues of how representative are the subpopulations inside the TME, and how large they should be, were discussed. Based on existing experience, an initial minimum of 100 individuals per TME for the key groups was advanced, but this number strongly depends on the selected organism group and the soil and/or ecosystem from which the TMEs are derived. Obviously, small-bodied organisms are usually more abundant in a TME core. For example, in the controls of TMEs sampled at 4 European sites, the number of enchytraeids in the top 5 cm of soil varied between 150 and 900 individuals, while the number of earthworms was 7 to 10 per TME core (Moser et al. 2004b; Römbke et al. 2004). For both groups, their number was clearly smaller in the TME from a Portuguese crop site than in those from Central or Northern European meadow sites. Therefore, better guidance on threshold values for minimum population sizes regarding their sustainability in the system and the possibility of a robust evaluation of effects is particularly required. Agreed was the need to have a pre-application incubation time (flexible, but varying from 1 to several weeks, and selected accordingly to expert knowledge on the ecology of the site), allowing communities to adapt and stabilize within the TME after the disturbance caused by the soil core extraction. This pre-application period is also important to allow sampling before the application, a relevant aspect in the use of data treatment strategies (e.g., BACI designs = before-after-control-impact (Figure 5.1)).

TMEs may potentially be applicable to different ecosystems, i.e., crop areas, grasslands, or even forests (Förster et al. 2006). However, existing experience shows a strong preference for grasslands due to the higher diversity and stability of soil organism communities, leading, in principle, to a lower variation in the initial composition of the TME, such as population size and community structure of main organism groups. Another reason behind the frequent selection of non-crop sites to collect TMEs is the perception of the lower sensitivity of in-field communities compared to off-field communities. In any case, the selection of the site should be ruled by the aim of the study, and research may be required to investigate the appropriateness of using grassland systems to reflect arable soils (in terms of both biota and pesticide exposure). Nevertheless, agreement was achieved on the need to select a homogeneous site (in terms of soil properties and vegetation cover) with a soil type being representative of the "typical soil" of the ecological region where the substance needs to be tested. In this respect, it will be necessary to define ecoregions, which are characterized by specific combinations of soil properties, climatic conditions, soil organism communities, and land use forms. Otherwise, it will be difficult to decide whether or not results from one TME study at one site can be extrapolated to other sites. In any case, extreme soils (very sandy or very clayey soils) should be avoided, mainly due to technical problems during the collection and maintenance of the TME. Regarding the collection of the TME, site edges should be avoided and, in order to avoid increasing the variability in the initial composition of TMEs, this operation should be conducted over a narrow area (Figure 5.2).

The size of a TME, regarding its representativeness of the soil system, was discussed, partly because the dimensions used so far were selected mainly due to

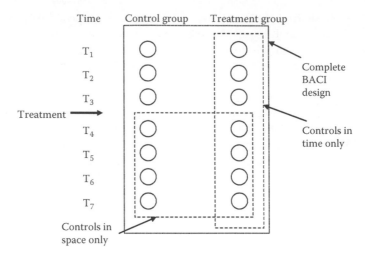

FIGURE 5.1   Figurative explanation of different possibilities regarding the design of a TME or field study. (Figure designed by Joost Lahr.)

practical considerations (originally, the tubes were made from gas storage devices, which were easily available and cheap). Thus, the selected size is a compromise solution between technical effort (including the possibility to deal with a higher number of replicates) and ecological relevance. In general, it is known that the use of small cores may lead to a biased representation of the system. Moreover, when collecting small-size cores, the probability of having a larger variance in the initial composition of the TME is higher, due to the grouped distribution of soil organism communities in the field. Size is also a function of the sampling strategy adopted, i.e., sub-sampling or destructive cores. While, in the past, usually smaller cores were used and sampled destructively, current research focuses on the question of the minimum size allowing sub-sampling (Roß-Nickoll, personal communication). Existing experience with TMEs supports sizes between 10 and 50 cm in diameter, and about 40 cm in depth. However the group felt that more studies are needed comparing the results from cores of different sizes.

The control of environmental conditions during the performance of a TME study was also addressed by the group. First, a comparison between indoor and outdoor systems was made. Both systems present advantages and disadvantages. With indoor systems, less realistic exposure conditions (almost like an extended laboratory test) are achieved. However, in this case, the closer control of environmental parameters, which may act as confounding factors when interpreting the data, is an advantage. This is particularly true for the soil moisture content in the TME. Evidence does exist that a differential water content level between different treatments (caused by different rates of evapotranspiration due to different plant biomass in the TME) can mask the effect of the test chemical on soil microbial activity parameters (Sousa et al. 2004). Moreover, a strong variation in soil microarthopods with each treatment was also reported as being caused by different soil water content levels (Koolhaas et al. 2004). This indicates, and the group agreed, that soil moisture content should

FIGURE 5.2 Extraction of TME soil cores from a central European meadow site. (Photo by Bernhard Förster.)

be measured and compensated for, regularly, by adopting a more intensive watering regime when required. New approaches include the use of soil sensors in order to address different moisture levels separately in different soil depths. On the other hand, with outdoor systems, more realistic exposure conditions are achieved, especially in relation to the natural seasonal fluctuations of the soil communities. However, the occurrence of extreme climatic conditions (especially drought events) may pose a problem when interpreting data and should be avoided. This is particularly true when this occurs with different intensities in different treatments, causing a biased interpretation of the results. Another problem posed by outdoor systems is the increased probability of a biased migration of individuals. The group agreed that more research on outdoor TME systems is clearly needed before their pros and cons can be adequately discussed.

The type of endpoints suitable for measurement in a TME study was intensively discussed in the group. Mesofauna community composition and structure was the effect parameter that the group felt to be particularly suited to this type of study, rather than functional parameters, although these should not be excluded. The measurement of species number, dominance structure, abundance, and trophic structure (e.g., in nematodes) can be appropriately assessed by a TME study. Several studies have shown that these parameters are suitable to be measured with this system (e.g., Moser et al. 2004a; Scholz-Starke et al. 2008; Kools et al. 2009). Besides measuring possible effects, it is also important to have the possibility to measure the intrinsic recovery of the system. Using an adequate sampling scheme (see next section), this can be achieved using TMEs coupled to the appropriate data treatment, in particular the use of multivariate methods (e.g., the PRC approach; see Section 5.4). However, depending on the substance to be tested, if recovery time takes 1 crop season, the use of indoor TMEs might be questioned. Usually, the duration of a TME study is 16 weeks, but being closed systems, they have a limited life span of up to 1 year (Römbke et al. 2004). However, longer-duration studies have not been tested so far.

It is possible to assess effects on soil earthworm communities (endogeic and anecic species), but this is constrained by several issues, namely, their density at the selected field and the possibility to have enough individuals within each TME. Although several studies showed the possibility of assessing effects on earthworm communities (Römbke et al. 2004; Förster et al. 2004), this parameter was considered optional by the group since there are probably many study sites (especially crop sites) where not enough earthworms occur to achieve sufficient numbers in soil cores. In addition, one must be aware that large-bodied individuals, especially deep-burrowing species such as *Lumbricus terrestris*, may (or may try to) escape the TME cores, since they usually crawl over the soil surface in search of food (Bouché 1976).

It is possible to assess functional endpoints in TME studies, and these show, in general, a lower variability than structural parameters. Soil collected at each TME can be used to measure microbial activity (e.g., soil enzymes), respiration, biomass, and diversity. Moreover, integrative functional parameters like litter decomposition (Förster et al. 2004, 2006) or the feeding rate via bait-lamina (Figure 5.3; Van Gestel et al. 2004) can also be measured within TMEs, although they may be too small to include litter bags. Similarly to soil fauna parameters, effects on these endpoints can also be evaluated using both univariate and multivariate methods. Nevertheless, the group felt that these parameters are optional depending on the aim of the study and the characteristics of the substance tested.

Although more research information is needed regarding the conduct and interpretation of effects data resulting from TME studies, some information is available showing the predictive value of the TME system regarding effects parameters. TME data available from an extensive study and a simultaneous field validation indicated that TMEs reflected not only effects, but also the natural variation encountered in the field for both soil microarthropods (Koolhaas et al. 2004; Moser et al. 2004b) and soil microbial parameters (Sousa et al. 2004).

Finally, the classification of the magnitude and duration of effects in TME studies was addressed by the group. The existing system for the aquatic compartment (Brock et al. 2000), the derived one adopted by Jänsch et al. (2006) for terrestrial systems, and that proposed in the Dutch report for the risk assessment of PPPs in soil (Van der Linden et al. 2006, 2008b) were discussed. The group considered the 2 existing classifications as promising tools to classify effects. However, a general opinion was that the threshold values adopted in terms of magnitude and duration of effects should be adapted taking into account the limitations of a TME study, in particular the variability between different samples and the power in detecting differences relating to the control (see Section 5.4), and also the maximum time period possible for a study (i.e., its suitability to detect recovery over long periods).

FIGURE 5.3 Conducting a bait-lamina test in a TME extracted from an agroforestry site. (Photo by Bernhard Förster.)

5.3 SAMPLING

First, the number and frequency of effect samplings were discussed in the relevant PERAS workgroup. It was obvious that the sampling design is mainly driven by the characteristics of the test substance, in particular its fate in soil. At the time of starting a semi-field study, such information should be available in detail from laboratory degradation and dissipation tests (e.g., OECD 2002), but often also from field dissipation studies. In particular, one has to differentiate in semi-field tests between what is required for nonpersistent and persistent chemicals. Some participants doubted whether there is a need for semi-field studies if the test substance is not persistent. However, rapid dissipation of the active ingredient does not automatically rule out that there is no soil metabolite that may be more persistent and/ or may still be able to affect soil organisms. In the case of nonpersistent test substances, sampling efforts should typically focus on the beginning of the study and potential recovery periods; i.e., frequently samples are taken in a geometric series.

In the case of persistent chemicals, just the opposite may be required, where the test substance is characterized by low toxicity and a potential for bioaccumulation, i.e., samples could be taken following a long exposure period. If the persistent test substance shows high toxicity, more resources are needed, since samples might be taken at both the beginning and the end of the study. Again, information is needed to decide which type of sampling is most suitable, and results from lower tiers should be used intensively.

Besides fate properties of the test substance, practical possibilities will define what sampling regime can be used. The number of samples as well as the total number of replicates will inevitably be limited. The design used in the earthworm field study may be a useful example to help in determining a sampling scheme (ISO 1998; Kula et al. 2006). According to experience gained with this regularly performed field test, samples should be taken before the start of the study and at least 4 times after application of the test substance for a period of up to 1 year. Using TMEs as an example, most semi-field studies performed so far have run for up to 16 weeks (e.g., Knacker et al. 2004), but individual studies prove that TME can be kept for longer, e.g., under glasshouse or laboratory conditions. For example, earthworm numbers and biomass were quite similar in terms of absolute numbers and seasonal patterns in TMEs and the field over 12 months (Römbke et al. 1994). In addition, preliminary experience from outdoor TME studies performed at the University of Aachen indicates that TME studies can be performed at least for 1 year (Roß-Nickoll, personal communication).

Such a long duration theoretically allows the study of recovery *sensu stricto*, i.e., an increase in population size after the test substance has negatively impacted the abundance or biomass of 1 or more species. However, when designing a TME study in order to include the recovery of the majority of impacted species, one must be aware that the duration of the life cycles of soil invertebrates can differ from a few weeks to about a year. An exception could be if the whole study focuses on one (or a few) very important species that are mainly affected by the test substance— as known from laboratory tests. However, such cases will probably be very rare since most ecologically important species (as far as we know) belong to the macro-fauna (e.g., ecosystem engineers such as earthworm vertical burrowers (Lavelle et al. 1997))—and such large-bodied species may be difficult to test in a TME with its limited size.

Summarizing the PERAS discussion about the minimum sampling efforts, it was recommended to take at least 1 pre-application sample and 4 post-application samples, divided into 2 samples for short-term effects (covering a range of 1 to 3 months after application) and 2 samples for recovery assessment, meaning that at least 6 are taken, and preferably 12 months after application. However, these recommendations should not be taken as a fixed scheme, since it is impossible to predict the duration of recovery of the soil organism community. Therefore, in order to be on the safe side, it may be worthwhile to design the study in a way that a fifth series of samples can be taken. Such a reserve may not always be needed, but when considering the resources needed to run a semi-field study, it is probably wise to have the possibility to take additional samples, since the alternative might be to require a whole new study.

It was generally agreed that, technically speaking, sampling does not differ considerably in TMEs from those methods recommended for field studies (e.g., ISO 2007a, 2007b, 2007c, 2007d). For example, samples for mesofauna are taken with a soil corer. In other cases, sample techniques must be modified. For example, the surface of a TME is usually not big enough to expose litter bags, meaning that decomposition of organic matter should be tested in another way. One possibility is to expose pieces of cellulose paper on the soil surface (Figure 5.4). Either mass loss or loss of area can be used as measurement endpoints (Förster et al. 2004).

However, in some cases certain soil properties may impede sampling efforts. One example is very sandy soils, from which it could be difficult to extract the soil core (the sand may be so loose that it cannot be taken out as an intact core). Also, very clayey soils may be too compacted for the extraction of intact soil cores. However, since it can be expected that TME studies will usually be performed with soils that can be utilized for agriculture (including meadows), such difficulties will be the exception, not the rule. This statement may be reconsidered when the study is going to be performed outside of the temperate regions of the world—but even with highly clayey soils of tropical lowland rain forests it has been possible to perform a TME study (Förster et al. 2006).

Finally, two issues that caused intensive discussion at PERAS could not be solved completely: Participants tried to clarify the pros and cons of taking several sub-samples from the same TME, or whether it is more suitable to sample the whole TME destructively. While the first approach is more efficient in terms of the use of resources, some people feared that, considering the small size of TMEs, the whole soil core is too greatly affected by the side effects of sampling. On the other hand, the "sacrifice" of whole soil cores at each sampling date means that many more soil cores have to be taken—and most of the sampled soil will be thrown away. The group agreed that more data are needed in order to decide which approach is recommended—or whether there are different recommendations, depending on the aim of the study and the endpoints selected. In fact, the same conclusion was reached for the question of whether soil cores should be kept in rooms with controlled conditions, or whether the TMEs should, at least partly, be exposed to field weather conditions (Figure 5.5). However, there was an overall agreement that environmental conditions should not unduly influence the study design, meaning that for indoor TMEs, standard conditions should ideally be defined for different regions.

5.4   STATISTICS

Variability within TME studies can be high due to the patchy distribution of soil organisms; however, they are probably no more variable than results obtained from field studies. Endpoints related to the structure of soil communities, such as abundance and diversity of organisms, are often more variable than functional endpoints, such as organic matter breakdown or the respiration rate. Some groups of organisms are naturally highly variable, e.g., oribatid mites with a coefficient of variation (CV) up to 70% or collembolans with a CV between 20% and 150%. It has been shown by a ring test that the variation of results between similarly performed studies is, however, quite comparable (Knacker et al., 2004).

FIGURE 5.4 Feeding activity on cellulose exposed horizontally on the soil surface (disks, rows 1 and 3) or vertically in the soil (rolls, rows 2 and 4) of treatments differing in carbendazim concentrations. (See Förster et al. (2004) for details; photo by Bernhard Förster.)

The variability due to natural heterogeneity in environmental conditions between soil cores taken from the field can be minimized by the careful choice of the test site and prescreening of the patchiness of organisms' distribution and soil properties. The smaller the plot from which the cores are sampled, the lower is the variability of the enclosed soil communities; it was discussed that samples should be taken from an area of about 5 × 5 m. Sub-sampling can be performed in the soil cores in order to

FIGURE 5.5 Outdoor experimental facility. The facility can hold up to 55 TME units that are stored on a well-drained soil-sand ground layer. (Photo by Björn Scholz-Starke.)

increase the replicate number and to reduce variability; however, it has to be shown that this technique is nondestructive and does not interfere with the communities in the residual core.

Before a semi-field study, such as a TME experiment, is started, a prospective power analysis for the test design based on already existing data on the variabilities may be of help in order to determine how many replicates are needed to achieve a desired minimal detectable difference (MDD). It was discussed that, for instance, 80% power to detect 50% deviation of a treated TME from a control TME could be a suitable threshold. Whether this assumption is achievable needs to be proven by experimental data.

Statistical analysis depends on the test design: From a dose-response experiment an EL/Cx of the test substance can be derived; if a NOEL/C has to be derived, the statistical power will strongly depend on the number of replicates. Until further experience has been gained by statistical power analysis of already performed TME studies, for practical reasons a basic design is proposed that uses at least 4 to 5 replicates per treatment, and at least 6 (preferably 8) replicates for the control treatment. When performing a limit test, with finite resources, more replicates are possible at the highest treatment rate to increase the power at that pesticide concentration.

Both uni- and multivariate statistical methods may be applied to help improve understanding and interpretation, and to determine the validity of the study. Faunistic data should be log-transformed before subsequent statistical treatment. The type of methods used will lead to various interpretation possibilities: Using univariate methods, the change in abundance of populations can be described; applying multivariate methods, e.g., principal response curves (PRCs), the alteration of community structure may be derived. NOEL/C values can be derived both for single populations and for the soil community. It was agreed that it is essential to integrate a suitable timeframe within the statistical design that is long enough to detect the recovery of affected communities.

It was agreed in the workshop that the whole data set of a semi-field test needs interpretation by experts with a sound knowledge of the biology and ecology of the soil biota.

6 Research Needs

In this chapter, research needs for semi-field test methods and in particular TME studies are summarized from discussions during the PERAS workshop. Focusing these needs on TME does not mean that research on other semi-field methods is not promoted (actually, many recommendations given in the following are true for various semi-field methods), but does reflect the discussions in the technical breakout groups.

Topics for basic ecological investigations are included, as well as research to better understand the uncertainty with regard to the extrapolation of experimental data from semi-field tests to the field and landscape scale. More specifically, basic research on the experimental setup is outlined with respect to the effect of soil properties, technical considerations of the TME systems, the application of the pesticide, the sampling strategy, statistics, and the limits of duration of semi-field tests.

6.1 PROTECTION GOALS AND RISK ASSESSMENT SCHEME

For the assessment of the acceptability of effects found in semi-field studies, a clear definition of soil protection goals is needed. From these protection goals, the level of protection can be deduced and, from this, the suitability of a test to show the magnitude and duration of certain effects can be determined. For acceptance of such an approach at a European level, PERAS proposed that a workshop should be organized to determine the appropriate protection goals for agricultural soil. Whether particular effects are acceptable over a certain time and place is, however, not a decision for scientists and risk assessors, but rather for risk managers. Therefore, risk managers and decision makers, in particular, should be invited.

The Dutch decision tree for persistent pesticides was described at PERAS as 1 national example that might stimulate discussion in this area. This scheme suggested 3 principles to set protection goals, that is, the functional redundancy principle (FRP), the community recovery principle (CRP), and the ecological threshold principle (ETP). These principles were applied for the in-crop situation, and the different principles were suggested for different time windows (see Section 3.3). In the Dutch proposal, the principles were elaborated into concrete decision schemes, including a tiered approach, but further research is needed to decide whether this approach can be translated to a spatial as well as temporal framework.

Note: Since the PERAS workshop, discussion of protection goals for soil is likely to be taken forward during revision of the European Commission terrestrial and persistence guidance documents.

6.2 BASIC ECOLOGICAL RESEARCH

Literature research and experimental work are necessary to describe the ecology, sensitivity, and recovery timeframes for mesofauna for both in-crop and off-crop communities. Also, the question of whether the use of life cycle traits is suitable to reflect the properties, e.g., the sensitivity of soil organism communities to PPPs, should be the subject of more detailed investigation. Therefore, quantitative descriptions of soil organism communities, including mean abundance and biomass of all important groups, and their distribution in space and time, reflected by their phenology and generation cycles, should be studied. In this respect, work should start with at least 2 groups of "soft-bodied" as well as "hard-bodied" organisms, i.e., enchytraeids, nematodes, collembolans, and mites.

The distribution of soil communities in the arable landscape is related to soil properties, vegetation pattern, and land use. These complex dependencies are not well characterized and need to be classified to understand the distribution patterns of the organisms. A valid classification system will then allow definition of reference situations suitable to integrate the landscape heterogeneity, such as different soil textures, pH, vegetation cover, and biotope types. Such a reference system could also be used to extrapolate to other site conditions and to compare in-crop vs. off-crop communities. Also, for extrapolation from one climatic region to another within Europe, a well-defined reference system for soil communities is necessary. A better understanding of soil organisms in the arable landscape can also help generate indicators for biodiversity.

In order to maximize efficiency when working with soil invertebrates, the development or, in the case of nematodes, the further improvement of expert systems for their taxonomic classification is recommended. Such methods, including genetical analysis of soil samples (again mainly for nematodes) will help to improve the routine identification of these organisms. In the meantime, i.e., before these expert systems are available, the preparation of taxonomic keys for reference systems, or specific regions, is a suitable approach. For example, keys for earthworms are available for Central and Northern Europe as well as for individual countries, such as Hungary, while all other regions are covered by individual papers at best. For mesofauna groups, the situation is usually worse, meaning that, at the least, compilations of existing literature have to be performed. Otherwise, an endpoint relating to the structure of soil organism communities cannot really be used.

6.3 UNCERTAINTIES IN EXTRAPOLATION

With respect to the risk assessment of pesticides, experimental data from lower-tier studies need to be related to, and validated against, environmental reality at the field and landscape levels. Therefore, experimental data from semi-field tests need to be extrapolated to other environmental conditions such as different soil types and temperature, water, and light regimes.

Several TME studies have been performed to test the effect of pesticides on soil organisms under various conditions, including incubation in the glasshouse or in the field, different types, concentrations, and application patterns of pesticides, monitoring of a range of organisms, different study durations, etc. However, no systematic research on the influence of different soil types on pesticide effects on soil biota in semi-field tests has yet been performed.

Keeping other conditions constant, the following soil properties should be specifically addressed to determine important correlations of abiotic soil properties with biocenoses: soil type and texture, the soil pH, the cation exchange and the water holding capacities, the organic matter content, and the nutrient status (C, N, P). The soil use and history should be included in such investigation. In particular, agricultural soils of different origins should be compared to nonagricultural soils, such as those from grassland.

Also, other soil parameters are of utmost importance for the fate and subsequently the effects of a biocide or pesticide on soil biota. Irrigation of the soil may be applied to either maintain constant soil moisture throughout the study or mimic the natural climatic conditions (rain).

Little is known about the variability of data from semi-field tests in comparison to data from field tests. Therefore, comparative experimental studies and models are considered necessary to determine the variability of data, for example, both within a TME with respect to sub-sampling and between independent TME cores, and to compare the variability of TME data with that of field test data. A literature database with regard to semi-field and field tests should be established based on already published data (Jänsch et al. 2006).

6.4 EXPERIMENTAL SETUP, SAMPLING, AND ANALYSIS

6.4.1 SOIL HYDROLOGY

Literature on microlysimeter experiments should be investigated to see if it will help determine how excavation may affect the soil hydrology.

Soil hydrology is of importance because it will influence the maintenance of the moisture content of the soil, and thus its biological activity, as well as the degradation and the transport of the pesticide.

The bottom of the soil column may be in contact with belowground soil or closed by a water-permeable inert material such as looped metal plates or porous ceramics. The different column closures will directly influence soil hydrology and therefore should be compared.

6.4.2 CORE SIZE, SIZE OF POPULATIONS, AND EQUILIBRATION

Research and guidance is needed with respect to the minimum soil core size and the appropriate sizes of soil populations in the cores that should be applied in TME studies.

The core sizes and the experimental setup in previous studies varied considerably: Sizes varied in diameter from 20 cm to 50 cm and in height from 40 cm to 60 cm, respectively. At smaller dimensions edge effects may occur. For a meaningful determination of pesticide effects on the population level, a critical minimum size of the different soil species in the soil column has to be defined.

The soil cores are usually sampled with their natural plant cover, transferred to the test site, and installed in the TME facilities. Before application of the pesticide the soil cores should reach an equilibrium that is defined by screening the abundance and distribution of organisms after installation of the cores. A minimum time period for TME to equilibrate has not been systematically investigated yet. Neither it is known under which conditions the cores should be kept during the equilibration period regarding watering, in case of dry periods, or covering them to protect from heavy rainfall if the cores are installed in the field. If the cores are installed in a glasshouse facility, the question of watering during the equilibration time is still of relevance.

6.4.3   APPLICATION

Application of a pesticide must consider the dissipation times of the active ingredient (and relevant soil metabolites), since with persistent pesticides an accumulation plateau will be established. Thus, the annual dose needs to be added on top of the baseline accumulation plateau in the top soil layer, e.g., the first 5 cm. However, it is not clear which mode of application would be most appropriate: The test substance may be applied at concentrations resembling both the accumulation plateau concentration and the annual dose simultaneously. If the soil cores have been sampled from sites with no previous application of the pesticide that will be investigated, the simultaneous application does not reflect the natural field situation because the accumulation plateau comprises, at least in part, aged residues usually with reduced bioavailability (Chung and Alexander, 2002). Upon simultaneous application, the soil organisms are exposed to high concentrations of freshly added pesticide with potentially toxic effects.

Alternatively, the pesticide may be added in a 2-step process: The first application will simulate the accumulation plateau concentration followed by an equilibration time simulating the aging process. Subsequently, after the equilibration time the annual dose of the pesticide may be applied. Although the aging period may be far shorter than in the real field experiment with applications from previous years, this strategy may more closely reflect the typical agricultural practice.

It remains to be investigated which application method should be used in semifield tests and what aging periods after the first pesticide application according to the accumulation plateau should be applied. The impact of incorporation on the soil biocenosis, and suitable recovery periods, should be considered case by case.

In addition, it has to be clarified whether exposure scenarios relevant for grassland sites can be extrapolated to crop sites. The background to this issue is the situation in earthworm field studies, which are often performed in grasslands due to the higher abundance and diversity of worms, while their results are used for risk assessment purposes at crop sites.

6.4.4   STATISTICS AND SAMPLING

Before a TME experiment is designed, a prospective power analysis for the test design should be performed in order to determine how many replicates are needed, for instance, to derive 80% power to detect 50% deviation of a treated TME from the control TME. Whether this assumption is achievable needs to be proven by experimental data.

Research in this respect should focus on 2 aspects: 1) How many independent replicates have to be installed in order to achieve a certain degree of probability to detect a level of difference in the treated TME compared to the untreated control TME? 2) It should be investigated whether subsequent sub-sampling of a TME at the defined sampling intervals, to increase the number of samples, will affect the soil communities of the residual soil columns compared to sacrificing individual soil cores at each sampling interval.

Further research may help to distinguish whether techniques like principal response curves are sufficient to explore community level effects, or whether diversity indexes can play an additional role.

6.4.5 STUDY DURATION

To test the recovery of soil communities after application of a pesticide, the duration of the test has to follow the fate and toxicity profile of the substance. Usually semi-field tests are designed to cover the growing season of the crop that is treated with a pesticide. If recovery of the organisms after crop harvest is not achieved, it may be necessary to extend the duration of the test over the next season, or the next year. So far, TME studies have been performed for a period of up to 1 year (Scholz-Starke et al. 2008). It remains to be tested whether studies can be set up for longer than 1 year to follow the long-term recovery of soil communities, where necessary.

6.4.6 SUMMARIZING AND EVALUATING

The most appropriate approach to compare different semi-field testing methods would be to establish experiments under similar conditions, i.e., the use of the same pesticide, dose, season, crop, soil, and other parameters that were discussed earlier. This represents the only scientific way to judge advantages and disadvantages of the test design and to assess which method might be most appropriate for the scientific problem to be solved.

A classification of the effects found, based on magnitude and duration of the effects, might be a good way forward to present the results in a concise way and to aid the interpretation of results from semi-field tests. It is considered that guidance needs to be developed to identify suitable triggers for semi-field studies and also for summarizing and evaluating experimental results in the context of their use in regulatory decision making. Such guidance has been and is being developed in the Netherlands for earthworms, aquatic model ecosystems, and non-target arthropods (see De Jong et al., 2006, 2008, 2009). It can be expected that for semi-field studies the same need exists. Therefore, it is recommended to identify suitable triggers for semi-field studies and to develop guidance for summarizing and evaluating the experimental results.

References

Akhouri NM, Kladivko EJ, Turco RF. 1997. Sorption and degradation of atrazine in middens formed by *Lumbricus terrestris*. Soil Biol Biochem 29:663–666.

Aldenberg T, Jaworska J. 2000. Uncertainty of hazardous concentrations and fraction affected for normal species sensitivity distributions. Ecotoxicol Environ Saf 46:1–18.

[ASTM] American Society for Testing and Materials. 1993. Standard guide for conducting a terrestrial soil-core microcosm test. Annual Book of ASTM 1197:546–557.

Axelsen JA, Holst N, Hamers T, Krogh PH. 1997. Simulations of the predator-prey interactions in a two species ecotoxicological test system. Ecol Model 101:15–25.

Bardgett RD. 2002. Causes and consequences of biological diversity in soil. Zoology 105:367–374.

Bardgett RD. 2007. The biology of soil. Oxford (UK): Oxford University Press.

Bardgett RD, Chan KF. 1999. Experimental evidence that soil fauna enhance nutrient mineralisation and plant nutrient uptake in montaine grassland ecosystems. Soil Biol Biochem 31:1007–1014.

Bauchhenss J. 1982. Artenspektrum, Biomasse, Diversität und Umsatzleistung von Lumbriciden (Regenwürmer) auf unterschiedlich bewirtschafteten Grünlandflächen verschiedener Standorte Bayerns. Bayer Landw Jb 59:119–124.

Bauchhenss J. 1997. Bodenzoologie. Boden-Dauerbeobachtungsflächen (BDF). Bericht nach zehnjähriger Laufzeit 1985–1995. Teil III. Boden: Gefüge, Organische Substanz, Bodenorganismen, Vegetation. Bodenkultur und Pflanzenbau. Schriftenr Bayer Landesamt Umweltschutz 6/97:219–234.

Bauer C, Römbke J. 1997. Factors influencing the toxicity of two pesticides on three lumbricid species in laboratory tests. Soil Biol Biochem 29:705–708.

Beck L. 1993. Zur Bedeutung der Bodentiere für den Stoffkreislauf in Wäldern. Biologie in unserer Zeit 23:286–294.

Beck L, Römbke J, Breure AM, Mulder C. 2005. Considerations for the use of ecological classification and assessment concepts in soil protection. Ecotoxicol Environ Safety 62:189–200.

Beck L, Römbke J, Ruf A, Prinzing A, Woas S. 2004. Effects of diflubenzuron and *Bacilus thuringiensis* var. *kurstaki* toxin on soil invertebrates of a mixed deciduous forest in the Upper Rhine Valley, Germany. Eur J Soil Biol 40:55–62.

Bembridge JD, Kedwards TJ, Edwards PJ. 1998. Variation in earthworm populations and methods for assessing responses to pertubations. In: Sheppard SC, Bembridge JD, Holmstrup M, Posthuma L, editors, Advances in earthworm ecotoxicology. Pensacola (FL): SETAC Press, p 341–352.

Binet F, Hallaire V, Curmi P. 1997. Agricultural practices and the spatial distribution of earthworms in maize fields. Relationships between earthworm abundance, maize plants and soil compaction. Soil Biol Biochem 29:577–583.

Boleas S, Alonso C, Pro J, Fernández C, Carbonell G, Tarazona JV. 2005. Toxicity of the antimicrobial oxytetracycline to soil organisms in a multi-species-soil system (MS•3) and influence of manure co-addition. J Hazard Mater 122:233–241.

Bouché MB. 1976. Etude de l'activite des invertebres epiges prairiaux. I. Resultats generaux et geodrilogiques (Lumbricidae: Oligochaeta). Revue Ecologie Biologie Sol 13:261–281.

Boyd JM. 1958. The ecology of earthworms in cattle-grazed machair in Tiree, Argyll. J Anim Ecol 27:147–157.

Bray JR, Curtis JT. 1957. An ordination of the upland forest communities of Southern Wisconsin. Ecol Monogr 27:325–349.

Briones MJI, Bol R. 2003. Natural abundance of 13C and 15N in earthworms from different cropping treatments: Earthworm isotopic values under different cropping treatments. Pedobiologia 47:560–567.

Brock TCM, Arts GHP, Maltby L, Van den Brink PJ. 2006. Aquatic risks of pesticides, ecological protection goals and common aims in EU legislation. Integrated Environ Assess Manage 2:e20–e46.

Brock TCM, Lahr J, Van den Brink PJ. 2000. Ecological risks of pesticides in freshwater ecosystems. Part 1. Herbicides. Technical Report 088.Wageningen (NE): Alterra Centre for Water and Climate.

Burrows LA, Edwards CA. 2002. The use of integrated soil microcosms to predict effects of pesticides on soil ecosystems. Eur J Soil Biol 38:245–249.

Burrows LA, Edwards CA. 2004. The use of integrated soil microcosms to assess the impact of carbendazim on soil ecosystems. Ecotoxicology 13:143–161.

Callahan CA, Menzie CA, Burmaster DE, Wilborn DC, Ernst T. 1991. On-site methods for assessing chemical impact on the soil environment using earthworms: a case study at the Baird and McGuire Superfund Site, Holbrook, Massachusetts. Environ Toxicol Chem 10:817–826.

Candolfi M, Blümel S, Forster R, Bakker FM, Grimm C, Hassan SA, Heimbach U, Mead-Briggs M, Reber B, Schmuck R, Vogt H. 2000. Guidelines to evaluate side-effects of plant protection products to non-target arthropods. Zurich (CH): International Organization for Biological and Integrated Control of Noxious Animals and Plants (IOBC/OILB).

Chapman PF, Maund SJ. 1996. Considerations for the experimental design of aquatic mesocosm and microcosm studies. In: Ostrander GK, editor, Techniques in aquatic toxicology. Boca Raton (FL): CRC Lewis Publishers, p 657–673.

Checkai R, Wentsel R, Phillips T, Yon RL. 1993. Controlled environment soil-core microcosm unit for investigating fate, migration, and transformation of chemicals in soils. J Soil Contam 2:229–243.

Christiansen K. 1964. Bionomics of Collembola. Annu Rev Entomol 9:147–178.

Chung N, Alexander M. 2002. Effect of soil properties on bioavailability and extractability of phenanthrene and atrazine sequestered in soil. Chemosphere 48:109–115.

Clarke KR. 1993. Non-parametric multivariate analysis of changes in community structure. Aust J Ecol 18:117–143.

Clarke KR, Warwick RM. 1994. Change in marine communities: an approach to statistical analysis and interpretation. 1st ed. Plymouth (UK): Plymouth Marine Laboratory.

Clarke KR, Warwick RM. 2001. Change in marine communities: an approach to statistical analysis and interpretation. 2nd ed. Plymouth (UK): PRIMER-E.

Clements RO, Murray PJ, Sturdy RG. 1991. The impact of 20 years' absence of earthworms and three levels of N fertilizer on a grassland soil environment. Agric Ecosyst Environ 36:75–85.

Coderre D, Maufette Y, Gagnon D, Tousignant S, Bessette G. 1995. Earthworm populations in healthy and declining sugar maple forests. Pedobiologia 39:86–96.

Cole LK, Metcalf RL, Sanborn JR. 1976. Environmental fate of insecticides in terrestrial model ecosystems. Int J Environ Stud 10:7–14.

Council Directive 67/548/EEC. 1967, June 27. On the approximation of laws, regulations, and administrative provisions relating to the classification, packaging, and labeling of dangerous substances.

Curry JP. 1998. Factors affecting earthworm abundance in soils. In: Edwards CA, editor, Earthworm ecology. Boca Raton (FL): CRC Press, p 37–64.

Czarnecki A, Losinski J. 1985. The effect of GT seed dressing on the community of Collembola in the soil under sugar beet. Pedobiologia 28:427–431.

Decaens T, Bureau F, Margerie P, Alard D. 2002. Earthworm communities in an agricultural wet landscape of the Seine valley (upper Normandy, France). 7th International Symposium on Earthworm Ecology, Cardiff (Wales), September 1–6, 2002. Book of Abstracts, p 11–12.

De Jong FMW, Bakker FM, Brown K, Jilesen CJTJ, Posthuma-Doodeman CJAM, Smit CE, van der Steen JJM. Forthcoming 2010. Guidance for summarising and evaluating field studies with non-target arthropods.

De Jong FMW, Brock TCM, Foekema EM, Leeuwangh P. 2008. Guidance for summarising aquatic micro- and mesocosm studies. A guidance document of the Dutch Platform for the Assessment of Higher Tier Studies. RIVM Report 601506009/2008. Bilthoven (NL): RIVM.

De Jong FMW, Mensink BJWG, Smit CE, Montforts MHMM. 2005. Evaluation and use of ecotoxicological field studies for regulatory purposes of plant protection products in Europe. HERA 11:1157–1176.

De Jong FMW, Van Beelen P, Smit CE, Montforts MHMM. 2006. Guidance for summarising earthworm field studies. A guidance document of the Dutch Platform for the Assessment of Higher Tier Studies. RIVM Report 601506006/2006. Bilthoven (NL): RIVM.

De Ruiter PC, Neutel A-M, Moore JC. 1998. Biodiversity in soil ecosystems: the role of energy flow and community stability. Appl Soil Ecol 10:217–228.

De Ruiter PC, Van Veen JA, Moore JC, Brussaard L, Hunt HW. 1993. Calculation of nitrogen mineralisation in soil food webs. Plant Soil 157:263–273.

De Vaufleury A, Kramarz PE, Binet P, Cortet J, Caul S, Andersen MN, Plumey E, Coeurdassier M, Krogh PH. 2007. Exposure and effects assessments of Bt-maize on non-target organisms (gastropods, microarthropods, mycorrhizal fungi) in microcosms. Pedobiologia 51:185–194.

Dietrich DR, Schmid P, Zweifel U, Schlatter C, Jenni-Eiermann S, Bachmann H, Buhler U, Zbinden N. 1995. Mortality of birds of prey following field application of granular carbofuran: a case study. Arch Environ Contam Toxicol 29:140–145.

Directive 98/8/EC of the European Parliament and of the Council. 1998, February 16. Concerning the placing of biocidal products on the market.

Directive 2000/60/EC of the European Parliament and of the Council. 2000, October 23. Establishing a framework for community action in the field of water policy.

Dunger W. 1983. Tiere im Boden. Wittenberg (DE): A. Ziemsen-Verlag.

Dykhuizen DE. 1998. Santa Rosalia revisited: why are there so many species of bacteria. Antonie van Leeuwenhoek 73:25–33.

[EC] European Commission. 2002, October 17. Guidance document on terrestrial ecotoxicology. Draft Working Document SANCO/10329/2002, rev. 2, final. Luxembourg (BE): European Commission.

[EC] European Commission. 2003. Technical guidance document on risk assessment. Part II. Luxembourg (BE): European Commission—Joint Research Centre, Institute for Health and Consumer Protection, European Chemicals Bureau, Office for Official Publications of the European Communities.

[EC] European Commission. 2009. Regulation (EC) No. 1107/2009 of the European Parliament and of the Council of 21 October 2009 concerning the placing of plant protection products on the market and repealing Council Directives 79/117/EEC and 91/414/EEC.

Edwards CA. 1983. Earthworm ecology in cultivated soils. In: Satchell JE, editor, Earthworm ecology—from Darwin to vermiculture. London: Chapman & Hall, p 123–137.

Edwards CA, Bohlen PJ. 1992. The effects of toxic chemicals on earthworms. Rev Environ Contam Toxicol 125:23–99.

Edwards CA, Bohlen PJ. 1996. Biology of earthworms. London: Chapman & Hall.

Edwards CA, Bohlen PJ, Linden DR, Subler S. 1995. Earthworms in agroecosystems. In: Hendrix PF, editor, Earthworm ecology and biogeography in North America. Boca Raton (FL): Lewis Publishing, p 185–215.

Edwards CA, Brown SM. 1982. Use of grassland plots to study the effect of pesticides on earthworms. Pedobiologia 24:145–150.

Edwards CA, Knacker TT, Pokarzhevskii AA, Subler S, Parmelee R. 1996. Use of soil micro-cosms in assessing the effects of pesticides on soil ecosystems. In: Proceedings of the International Symposium on Environmental Behavior of Crop Protection Chemicals. IAEA-SM-343/3:435-352. Vienna (AT): International Atomic Energy Agency.

[EFSA] European Food Safety Authority. 2007. Opinion of the Scientific Panel on Plant Protection Products and Their Residues on a request from the commission related to the revision of Annexes II and III to Council Directive 91/414/EEC concerning the placing of plant protection products on the market—ecotoxicological studies. EFSA J 461:1–44.

Ehrmann O. 1996. Regenwürmer in einigen südwestdeutschen Agrarlandschaften: Vorkommen, Entwicklung bei Nutzungsänderungen und Auswirkungen auf das Bodengefüge. Hohenh Bodenk H 35:1–135.

[EMEA] European Medicines Agency. 2007, May 2. Guideline on environmental impact assessment for veterinary medicinal products in support of the VICH guidelines GL6 and GL38. EMEA/CVMP/ERA/418282/2005. London: European Medicines Agency, Committee for Medicinal Products for Veterinary Use (CVMP).

[EPPO] European and Mediterranean Plant Protection Organization. 2003. EPPO standards, PP 3/7 (revised): environmental risk assessment scheme for plant protection products. EPPO Bull 33:147–149.

European Council Directive 91/414/EEC. 1991, July 15. Concerning the placing of plant protection products on the market.

Evans AC, Guild WJMC. 1947. Some notes on reproduction in British earthworms. Ann Mag Nat Hist 654.

Fernandez C, Alonso C, Babin MM, Pro J, Carbonell C, Tarazona JV. 2004. Ecotoxicological assessment of doxycycline in aged pig manure using multispecies soil systems (MS•3). Sci Total Environ 323:63–69.

Filser J. 1992. Dynamik der Collembolengesellschaften als Indikatoren für bewirtschaftungsbedingte Bodenbelastungen. Hopfenböden als Beispiel. Dissertation, Ludwig-Maximilian-Universität München.

Filser J. 1995. The effect of green manure on the distribution of Collembola in a permanent row crop. Biol Fertil Soils 19:303–308.

FOCUS. 2000. FOCUS groundwater scenarios in the EU review of active substances. Report of the FOCUS Groundwater Scenarios Workgroup, EC Document SANCO/321/2000, rev. 2.

Förster B, Garcia MVB, Francimari O, Römbke J. 2006. Effects of carbendazim and lambda-cyhalothrin on soil invertebrates and leaf litter decomposition in semi-field and field tests under tropical conditions (Amazonia, Brazil). Eur J Soil Biol 42:S171–S179.

Förster B, Van Gestel CAM, Koolhaas JEE, Nentwig G, Rodrigues JML, Sousa JP, Joes SE, Knacker T. 2004. Ring-testing and field-validation of a terrestrial model ecosystem (TME)—an instrument for testing potentially harmful substances: effects of carbendazim on organic matter breakdown and soil fauna feeding activity. Ecotoxicology 13:129–141.

Frampton GK. 1999. Spatial variation in non-target effects of the insecticides chlorpyrifos, cypermethrin and pirimicarb on Collembola in winter wheat. Pesticide Sci 55:875–886.

Frampton GK. 2001. The effects of pesticide regimes on non-target arthropods. In: Joung JEB, Griffin MJ, Alford DV, Ogilvy SE, editors, Reducing agrochemical use on the arable farm. London: DEFRA, p 219–214.

Frampton GK. 2007. Collembola and macroarthropod community responses to carbamate, organophosphate and synthetic pyrethroid insecticides: direct and indirect effects. Environ Pollut 147:14–25.

Frampton GK, Jaensch S, Scott-Fordsmand JJ, Roembke J, van den Brink PJ. 2006. Effects of pesticides on soil invertebrates in laboratory studies: a review and analysis using species sensitivity distributions. Environ Toxicol Chem 25:2480–2489.

Frampton GK, Jones SE, Knacker T, Förster B, Römbke J, Filser J, Mebes H. 2002. Assessing the effects of pesticides on non-target soil organisms involved in the degradation of organic matter. Unpublished Research and Development Final Project Report PN0938. London: DEFRA. Available from http://randd.defra.gov.uk/.

Frampton GK, van den Brink PJ, Gould PJL. 2000. Effects of spring precipitation on a temperate arable collembolan community analysed using principal response curves. Appl Soil Ecol 14:231–248.

Frampton GK, van den Brink PJ, Wratten SD 2001. Diel activity patterns in an arable collembolan community. Appl Soil Ecol 17:63–80.

Frampton GK, Wratten SD. 2000. Effects of benzimidazole and triazole fungicide use on epigeic species of Collembola in wheat. Ecotoxicol Environ Safety 46:64–72.

Fraser LH, Keddy P. 1997. The role of experimental microcosms in ecological research. Trends Ecol Evol 12:478–481.

Führ F, Hance RJ, editors. 1992. Lysimeter studies of the fate of pesticides in the soil. British Crop Protection Council Monograph 53:1–192. Lavenham, Suffolk (UK): Lavenham Press Ltd.

Gerard BM. 1967. Factors affecting earthworms in pastures. J Anim Ecol 36:235–252.

Giddings JM, Brock TCM, Heger W, Heimbach F, Maund SJ, Norman SM, Ratte HT, Schäfers C, Streloke M. 2002. Community-level aquatic system studies—interpretation criteria. Proceedings from the CLASSIC Workshop, Schmallenberg, Germany. Pensacola (FL): SETAC Press.

Gile JD. 1983. 2,4-D—its distribution and effects in a ryegrass ecosystem. J Environ Qual 12:406–412.

Gile JD, Collins JC, Gillett JW. 1980. Fate of selected herbicides in a terrestrial laboratory microcosm. Environ Sci Technol 14:1124–1128.

Gillett JW, Gile JD. 1976. Pesticide fate in terrestrial laboratory ecosystems. Int J Environ Stud 10:15–22.

Gisin H. 1943. Ökologie und Lebensgemeinschaften der Collembolen im Schweizer Exkursionsgebiet Basel. Rev Ecol Biol Sol 50:131–224.

Graff O. 1953. Die Regenwürmer Deutschlands. Schriftenreihe der Forschungsanstalt für Landwirtschaft Braunschweig-Völkenrode 7:1–70.

Griffiths BS, Ritz K, Bardgett RD, Cook R, Christensen S, Ekelund F, Sørensen SJ, Bååth E, Bloem J, De Ruiter PC, Dolfing J, Nicolardot B. 2000. Ecosystem response of pasture soil communities to fumigation-induced microbial diversity reductions: an examination of the biodiversity-ecosystem function relationship. Oikos 90:279–294.

Heimann-Detlefsen D. 1991. Auswirkungen eines unterschiedlich intensiven Pflanzenschutz- und Düngemitteleinsatzes auf die Collembolenfauna des Ackerbodens. Dissertation, Uni Braunschweig.

Heimbach F. 1985. Comparison of laboratory methods, using *Eisenia foetida* and *Lumbricus terrestris*, for the assessment of the hazard of chemicals to earthworms. Z Pflanzenk Pflanzen 92:186–193.

Heimbach U, Dohmen P, Barrett KL, Jäckel B, Kennedy PJ, Mead-Briggs M, Nienstedt KM, Römbke J, Schmitzer S, Schmuck R, Ufer A, Wilhelmy H. 2000. A method for testing effects of plant protection products on the carabid beetle (Coleoptera: Carabidae) under laboratory and semi-field conditions. In: Candolfi MP, Blümel S, Forster R, et al., editors, Guidelines to evaluate side-effects of plant protection products to non-target arthropods. Gent (BE): IOBC Publishers, p 87–106.

Högger CH. 1994. Überprüfung der Regenwurm-Toxizität von Pflanzenbehandlungsmitteln. Unveröff. Manuskript, FAP Zürich.

Hoogerkamp M. 1987. Effect of earthworms on the productivity of grassland; an evaluation. In: Bonvicini Pagliai AM, Omodeo P, editors, On earthworms. Modena (IT): Mucchi Editore, p 485–495.

Hopkin SP. 1997. Biology of the springtails (Insecta: Collembola). Oxford (UK): Oxford University Press.

Hunt HW, Coleman DC, Ingham ER, Ingham RE, Elliott ET, Moore JC, Rose SL, Reid CPP, Morley CR. 1987. The detrital food web in a shortgrass prairie. Biol Fertil Soils 3:57–68.

Inglesfield C. 1984. Toxicity of the pyrethroid insecticides Cypermethrin and WL85871 to the earthworm *Eisenia foetida* Savigny. Bull Environ Contam Toxicol 33:568–570.

[ISO] International Organisation for Standardisation. 1998. Soil quality—effects of pollutants on earthworms (*Eisenia fetida*). Part 3. Guidance on the determination of effects in field situations. ISO 11268-2. Paris (FR): ISO.

[ISO] International Organization for Standardization. 2007a. Soil quality—sampling of soil invertebrates. Part 1. Hand-sorting and formalin extraction of earthworms. ISO 23611-1. Geneva (CH): ISO.

[ISO] International Organization for Standardization. 2007b. Soil quality—sampling of soil invertebrates. Part 2. Sampling and extraction of microarthropods (Collembola and Acarina). ISO 23611-2. Geneva (CH): ISO.

[ISO] International Organization for Standardization. 2007c. Soil quality—sampling of soil invertebrates. Part 3. Sampling and soil extraction of enchytraeids. ISO 23611-3. Geneva (CH): ISO.

[ISO] International Organization for Standardization. 2007d. Soil quality—sampling of soil invertebrates. Part 4. Sampling, extraction and identification of free-living stages of nematodes. ISO 23611-4. Geneva (CH): ISO.

Jänsch S, Frampton GK, Roembke J, van den Brink PJ, Scott-Fordsmand JJ. 2006. Effects of pesticides on soil invertebrates in model ecosystem and field studies: a review and comparison with laboratory toxicity data. Environ Toxicol Chem 25:2490–2501.

Jones A, Hart ADM. 1998. Comparison of laboratory toxicity tests for pesticides with field effects on earthworm populations: a review. In: Sheppard SC, Bembridge JD, Holmstrup M, Posthuma L, editors, Advances in earthworm ecotoxicology. Pensacola (FL): SETAC Press, p 247–267.

Kampichler C, Bruckner A, Baumgarten A, Berthold A, Zechmeister-Boltenstern S. 1999. Field mesocosms for assessing biotic processes in soils: how to avoid side effects. Eur J Soil Biol 35:135–143.

Kampichler C, Bruckner A, Kandeler E, Bauer R, Wright J. 1995. A mesocosm study design using undisturbed soil monoliths. Acta Zool Fenn 196:71–72.

Karlen DL, Mausbach MJ, Doran JW, Cline RG, Harris JF, Schuman GE. 1997. Soil quality: a concept, definition, and framework for evaluation. Soil Sci Soc Am J 61:4–10.

Kennel W. 1990. The role of the earthworm *Lumbricus terrestris* in integrated fruit production. Acta Horticult 285:149–156.

Knacker T, Van Gestel CAM, Jones SE, Soares AMVM, Schallnaß H-J, Förster B, Edwards CA. 2004. Ring-testing and field validation of a terrestrial model ecosystem (TME)—an instrument for testing potentially harmful substances: conceptual approach and study design. Ecotoxicology 13:5–23.

Koolhaas JE, Van Gestel CAM, Römbke J, Soares AMVM, Jones SE. 2004. Ring-testing and field-validation of a terrestrial model ecosystem (TME)—an instrument for testing potentially harmful substances: effects of carbendazim on soil microarthropod communities. Ecotoxicology 13:75–88.

Kools SAE. 2006. Soil ecosystem toxicology. Metal effects on structure and function. PhD thesis, Vrije Universiteit, Amsterdam (NL).

Kools SAE, Boivin MEY, Van Der Wurff AWG, Berg MP, Van Gestel CAM, Van Straalen NM. 2009. Assessment of structure and function in metal polluted grasslands using terrestrial model ecosystems. Ecotoxicol Environ Safety 72:51–59.

Kühle JC. 1986. Modelluntersuchungen zur strukturellen und ökotoxikologischen Belastung von Regenwürmern in Weinbergen Mitteleuropas (Oligochaeta: Lumbricidae). Bonn (DE): Rheinische Friedrich-Wilhelms-Universität.

Kula C, Heimbach F, Riepert F, Römbke J. 2006. Technical recommendations for the update of the ISO Earthworm Field Test Guideline (ISO 11268-3). J Soils Sediment 6:182–186.

Kuperman R, Knacker T, Checkai R, Edwards C. 2002. Multispecies and multiprocess assays to assess the effects of chemicals on contaminated sites. In: Sunahara GI, et al., editors, Environmental analysis of contaminated sites. Chichester (UK): John Wiley & Sons, p 45–60.

Laakso J, Setälä H. 1999. Sensitivity of primary production to changes in the architecture of belowground food webs. Oikos 87:57–64.

Lardier PA, Schiavon M. 1989. Toxicite du carbofuran et activite synergique de l'Atrazine sur son action vis-a-vis de quelques especes animales. Agronomie 9:959–963.

Lavelle P, Bignell D, Lepage M. 1997. Soil function in a changing world: the role of invertebrate ecosystem engineers. Eur J Soil Biol 33:159–193.

Lee KE. 1985. Earthworms—their ecology and relationships with soils and land use. Maryland Heights (MD): Academic Press.

Liber K, Kaushik NK, Solomon KR, Carey JH. 1992. Experimental designs for aquatic mesocosm studies: a comparison of the "ANOVA" and "regression" design for assessing the impact of tetrachlorophenol on zooplankton populations in limnocorrals. Environ Toxicol Chem 11:61–77.

Liess M, Brown C, Duquesne S, Dohmen P, Hart A, Heimbach F, Kreuger J, Lagadic L, Reinert W, Maund S, Streloke M, Tarazona J, editors. 2005. Effects of pesticides in the field (EPIF). Pensacola (FL): SETAC Publ.

Lofs-Holmin A. 1982. Influence of routine pesticide spraying on earthworms (Lumbricidae) in field experiments with winter wheat. Swed J Agric Res 12:121–123.

Løkke H. 1995. Effects of pesticides on meso- and microfauna in soil. Report 8. Københaun (DK): Danish Environmental Protection Agency.

Lübben B. 1991. Auswirkungen von Klärschlammdüngung und Schwermetallbelastung auf die Collembolenfauna eines Ackerbodens. Dissertation, Uni Braunschweig.

Metcalf RL, Sangha GK, Inder PK. 1971. Model ecosystem for the evaluation of pesticide biodegradability and ecological magnification. Environ Sci Technol 5:709–713.

Metge K, Heimbach U. 1998. Tests on the staphylinid *Philonthus cognatus*. In: Løkke H, Van Gestel CAM, editors, Handbook of soil invertebrate toxicity tests. Chichester (UK): Wiley & Sons, p 157–179.

Morgan E, Knacker T. 1994. The role of laboratory terrestrial model ecosystems in the testing of potentially harmful substances. Ecotoxicology 3:213–233.

Moser T, Römbke J, Schallnass H-J, Van Gestel CAM. 2007. The use of the multivariate principal response curve (PRC) for community level analysis: a case study on the effects of carbendazim on enchytraeids in terrestrial model ecosystems (TME). Ecotoxicology 16:573–583.

Moser T, Schallnass HJ, Jones SE, Van Gestel CAM, Koolhaas JE, Rodrigues JML, Römbke J. 2004a. Ring-testing and field-validation of a terrestrial model ecosystem (TME)—an instrument for testing potentially harmful substances: effects of carbendazim on nematodes. Ecotoxicology 13:61–74.

Moser T, Van Gestel CAM, Jones SE, Koolhaas JE, Rodrigues JML, Römbke J. 2004b. Ring-testing and field-validation of a terrestrial model ecosystem (TME)—an instrument for testing potentially harmful substances: effects of carbendazim on enchytraeids. Ecotoxicology 13:89–103.

Mucina L, Grabherr G, Ellmauer T. 1993. Die Pflanzengesellschaften Österreichs. Teil 1. Anthropogene Vegetation. Jena (DE): Gustav Fischer.

[NAFTA] North American Free Trade Agreement. 2005. Guidance document for conducting terrestrial field dissipation studies. USEPA (Washington [DC]: Office of Pesticide Programs) and Health Canada (Ottowa [CN]: Pest Management Regulatory Agency).

Nahmani J, Lavelle P, Rosi J-P. 2006. Does changing the taxonomical resolution alter the value of soil macroinvertebrates as bioindicators of metal pollution? Soil Biol Biochem 38:385–396.

Odum EP. 1984. The mesocosm. Bioscience 34:558–562.

Odum EP. 1985. Fundamentals in ecology. 3rd ed. Philadelphia (PA): Saunders College Publishing.

Odum HT. 1971. Environment, power and society. New York: Wiley.

[OECD] Organization for Economic Cooperation and Development. 1984. Earthworm, acute toxicity tests. Guideline for Testing Chemicals 207:16–21. Paris: OECD.

[OECD] Organization for Economic Cooperation and Development. 2002. Aerobic and anaerobic transformation in soil. Guideline for Testing Chemicals 307:16–21. Paris: OECD.

Peachey JE. 1963. Studies on the Enchytraeidae (Oligochaeta) of moorland soil. Pedobiologia 2:81–95.

Pernin C, Ambrosi JP, Cortet J, Joffre R, Le Petit J, Tabone E, Torre F, Krogh PH. 2006. Effects of sewage sludge and copper enrichment on both soil mesofauna community and decomposition of oak leaves (Quercus suber) in a mesocosm. Biol Fertil Soils 43:39–50.

Peters D. 1984. Die Regenwurmfauna verschieden genutzter Böden im Raum Krefeld (Lumbricidae). Decheniana 138:118–134.

Petersen H, Luxton M. 1982. A comparative analysis of soil fauna populations and their role in decomposition processes. Oikos 39:287–388.

PflSchG. 1998. Gesetz zum Schutz der Kulturpflanzen (Pflanzenschutzgesetz—PflSchG). Bundesgesetzblatt, No. 28, May 27.

Pignatti S. 1994. A complex approach to phytosociology. Ann Bot (Rome) 52:65–80.

Posthuma L, Suter GW, Traas TP, editors. 2002. The use of species sensitivity distributions in ecotoxicology. Boca Raton (FL): Lewis Publishers.

Rapport DJ, Costanza R, McMichael AJ. 1998. Assessing ecosystem health. Trends Ecol Evol 13:397–402.

Raw F. 1962. Studies of earthworm populations in orchards. I. Leaf burial in apple orchards. Ann Appl Biol 50:389.

Regulation (EC) 1907/2006. 2006, December 18. Of the European Parliament and of the council, concerning the Registration, Evaluation, Authorisation and Restriction of Chemicals (REACH), establishing a European chemicals agency, amending Directive 1999/45/EC and repealing Council Regulation (EEC) 793/93 and Commission Regulation (EC) 1488/94, as well as Council Directive 76/769/EEC and Commission Directives 91/155/EEC, 93/67/EEC, 93/105/EC, and 2000/21/EC.

Römbke J, Bauer C, Marschner A. 1996. Hazard assessment of chemicals in soil. Proposed ecotoxicological test strategy. Environ Sci Pollut Res 3:78–82.

Römbke J, Beck L, Dreher P, Hund-Rinke K, Jänsch S, Kratz W, Pieper S, Ruf A, Spelda J, Woas S. 2002. Entwicklung von bodenbiologischen Bodengüteklassen für Acker- und Grünlandstandorte. Berlin (DE): UBA-Texte 20/02.

Römbke J, Beck L, Förster B, Fründ C-H, Horak F, Ruf A, Rosciczweski K, Scheurig M, Woas S. 1997. Boden als Lebensraum für Bodenorganismen und bodenbiologische Standortklassifikation—Literaturstudie. Texte und Berichte zum Bodenschutz 4/97. Karlsruhe (DE): Landesanstalt für Umweltschutz Baden-Württemberg.

Römbke J, Breure AM, editors. 2005. Ecological soil quality—classification and assessment. Ecotoxicol Environ Safety 62:185–308.

Römbke J, Heimbach F, Hoy S, Kula C, Scott-Fordsmand J, Sousa P, Stephenson G, Weeks J. 2003. Effects of Plant Protection Products on Functional Endpoints in Soils (EPFES): Lisbon, 24–26 April 2002. Pensacola (FL): SETAC Press.

Römbke J, Jones SE, Van Gestel CAM, Koolhaas J, Rodrigues J, Moser T. 2004. Ring-testing and field-validation of a terrestrial model ecosystem (TME)—an instrument for testing potentially harmful substances: effects of carbendazim on earthworms. Ecotoxicology 13:105–118.

Römbke J, Knacker Th, Förster B, Marcinkowski A. 1994. Comparison of effects of two pesticides on soil organisms in laboratory tests, microcosms and in the field. In: Donker M, Eijsackers H, Heimbach F, editors, Ecotoxicology of soil organisms. Chelsea (MI): Lewis Publ., p 229–240.

Römbke J, Moltmann JF. 1996. Applied ecotoxicology. Boca Raton (FL): CRC Press.

Rundgren S. 1975. Vertical distribution of lumbricids in southern Sweden. Oikos 26:299–306.

Ruppel RF, Laughlin CW. 1976. Toxicity of some soil pesticides to earthworms. J Kans Entomol Soc 50:113–118.

Satchell JE. 1955. Some aspects of earthworm ecology. Soil Zool (Kevan) 180–201.

Satchell JE, editor. 1983. Earthworm ecology—from Darwin to vermiculture. London: Chapman & Hall.

Schäfer M, Tischler W. 1983. Ökologie. Jena (DE): Gustav Fischer Verlag.

Scheu S. 2001. Plants and generalist predators as links between the below-ground and above-ground system. Basic Appl Ecol 3:3–13.

Scholz-Starke B, Theissen B, Nikolakis A, Schäffer A, Roß-Nickoll M. 2008. Ecological and ecotoxicological evaluation of terrestrial model ecosystems (TME). Warsaw (PL): SETAC.

Schuphan I. 1986. Determination of the quantitive ecochemical and ecotoxicological behaviour of pesticides using vegetation chambers with controlled ventilation conditions. Plant Res Dev 23:92–108.

Schuphan I, Schärer E, Heise M, Ebing W. 1987. Use of laboratory model ecosystems to evaluate quantitatively the behaviour of chemicals. In: Greenhalgh R, Roberts TR, editors, Pesticide science and biotechnology. Oxford (UK): Blackwell Scientific Publishing, p 437–444.

Scott-Fordsmand JJ, Maraldo K, Van den Brink P. 2008. The toxicity of copper contaminated soil using a gnotobiotic soil multi-species test system (SMS). Environ Int 34:524–530.

Sheppard SC. 1997. Toxicity testing using microcosms. In: Tarradellas J, Bitton G, Rossel D, editors, Soil ecotoxicology. Boca Raton (FL): Lewis Publishing, p 345–373.

Smit CE, Schouten AJ, Van den Brink PJ, Van Esbroek MLP, Posthuma L. 2002. Effects of zinc contamination on a natural nematode community in outdoor soil mesocosms. Arch Environ Contam Toxicol 42:205–216.

Sousa JP, Rodrigues JML, Loureiro S, Soares AMVM, Jones SE, Förster B, Van Gestel CAM. 2004. Ring-testing and field-validation of a terrestrial model ecosystem (TME)—an instrument for testing potentially harmful substances: effects of carbendazim on soil microbial parameters. Ecotoxicology 13:43–60.

Stähli R, Suter E, Cuendet G. 1997. Die Regenwurmfauna von Dauergrünland des Schweizer Mittellandes. Bern (CH): BUWAL-Schriftenreihe Umwelt 291.

Sterzynska M. 1990. Communities of Collembola in natural and transformed soils of the linden-oak-hornbeam sites of the Mazovian lowland. Fragmenta Faunistica 34:166–260.

Stockdill SMJ. 1982. Effects of introduced earthworms on the productivity of New Zealand pastures. Pedobiologia 24:29–35.

Swift MJ, Heal OW, Anderson JM. 1979. Decomposition in terrestrial ecosystems. Oxford (UK): Blackwell Scientific Publications.

Theißen B. 2009. Klassifizierung ausgewählter Boden-Mesofauna-Taxozönosen (Collembola und Gamasina) von Ackerrandstreifen. Dissertation, RWTH Aachen University.

Todd TC, James SW, Seastedt TR. 1992. Soil invertebrate and plant responses to mowing and carbofuran application in a North American tallgrass prairie. Plant Soil 144:117–124.

Torsvik V, Goksøyr J, Daae FL. 1990. High diversity of DNA of soil bacteria. Appl Environ Microbiol 56:782–787.

Tscharntke T, Hawkins BA. 2001. Multitrophic interactions in terrestrial systems. Cambridge (UK): Cambridge University Press.

[UBA] Umweltbundesamt. 1994. UBA—workshop on terrestrial model ecosystems. Berlin: UBA-Texte 54/94.

Ufer A. 1993. Regeneration von Agrarbiotopen nach Bodenentseuchungs- und Pflanzenschutzmaßnahmen am Beispiel von Collembolen und Spinnen. Dissertation, Uni Heidelberg.

[USEPA] US Envoronmental Protection Agency. 1996. Ecological effects test guidelines. OPPTS 850.2450. Terrestrial (soil-core) microcosm test. Public draft. Washington (DC): USEPA.

Van den Brink PJ, Ter Braak CJF. 1998. Multivariate analysis of stress in experimental ecosystems by principal response curves and similarity analysis. Aquat Ecol 32:163–178.

Van den Brink PJ, Ter Braak CJF. 1999. Principal response curves: analysis of time-dependent multivariate responses of biological community to stress. Environ Toxicol Chem 18:138–148.

Van der Linden AMA, Boesten JJTI, Brock TCM, Van Eekelen GMA, De Jong FMW, Leistra M, Montforts MHMM, Pol JW. 2006. Persistence of plant protection products in soil; a proposal for risk assessment. RIVM Report 601506008/2006. Bilthoven (NL): RIVM. Available from: www.rivm.nl/bibliotheek/rapporten/601506008.pdf.

Van der Linden AMA, Boesten JJTI, Brock TCM, Van Eekelen GMA, Ter Horst MMS, De Jong FMW, Montforts MHMM, Pol JW. 2008a. Evaluation of the 2006 proposal for risk assessment of persistence of plant protection products in soil. RIVM Report 601712002. Bilthoven (NL): RIVM.

Van der Linden AMA, Boesten JJTI, Brock TCM, Van Eekelen GMA, Ter Horst MMS, De Jong FMW, Montforts MHMM, Pol JW. 2008b. Revised proposal for the risk assessment of persistence of plant protection products in soil. RIVM Report 601712003/2008. Bilthoven (NL): RIVM.

Van Gestel CAM, Koolhaas JEE, Schallnass H-J, Rodrigues JML, Jones SE. 2004. Ring-testing and field-validation of a terrestrial model ecosystem (TME)—an instrument for testing potentially harmful substances: Effects of carbendazim on nutrient cycling. Ecotoxicology 13:119–128.

Van Voris P, Tolle DA, Arthur MF. 1985. Experimental terrestrial soil-core microcosm test protocol. EPA/600/3-85/047 PNL-5450, UC-11. Washington (DC): USEPA.

Van Wensem J, Jagers op Akkerhuis GAJM, Van Straalen NM. 1991. Effect of the fungicide triphenyltin hydroxide on soil mediated litter decomposition. Pestic Sci 32:307–316.

Vera FWM. 2000. Grazing ecology and forest history. Wallingford (UK): CAB–International.

Verhoef HA. 1996. The role of soil microcosms in the study of ecosystem processes. Ecology 77:685–690.

Wall DH, Adams G, Parsons AN. 2001. Soil biodiversity. In: Chapin FS III, Sala OE, Huber-Sannwald E, editors, Global biodiversity in a changing environment. Scenarios for the 21st century. Ecol Stud 152:47–82.

Weidemann G. 1986. Artenzahlen in den Ökosystemen des Sollings. In: Ellenberg H, Mayer R, Scheuermann J, editors, Ökosystemforschung—Ergebnisse des Solling-Projekts 1966–1986. Stuttgart (DE): Verlag Eugen Ulmer, p 219–225.

Weyers A, Sokull-Klüttgen B, Knacker T, Martin S, Van Gestel CAM. 2004. Use of terrestrial model ecosystem data in environmental risk assessment for industrial chemicals, biocides and plant protection products in the EU. Ecotoxicology 13:163–176.

Wiles JA, Jepson PC. 1992. In-situ bioassay techniques to evaluate the toxicity of pesticides to beneficial invertebrates in cereals. Aspects Appl Biol 31:61–68.

Zechmeister HG, Moser D. 2001. The influence of agricultural land-use intensity on bryophyte species richness. Biodivers Conserv 10:1609–1625.

Appendix 1: Workshop Program

October 8, 2007

08.30–08.45	Welcome and introduction, aims of the workshop
08.45–09.15	Regulatory and industry views and expectations
09.15–09.45	Ecological background and context
09.45–10.15	Regulatory context: Directive 91/414/EEC and future pesticide regulation
10.15–10.45	Soil risk assessment of persistent pesticides: Dutch proposal and other frameworks
10.45–11.00	Discussion
11.30–12.15	Overview and evaluation of soil higher-tier methods
12.15–12.30	Discussion
14.00–14.15	Breakout group instruction
14.15–17.00	First breakout group session: Discussion of different approaches and their regulatory use
	Group 1: Pros and cons of semi-field vs. laboratory single- or multispecies approaches
	Group 2: Pros and cons of semi-field vs. full-scale field methods
	Group 3: Regulatory use of semi-field methods in tiered testing procedure
17.00–18.00	First plenary session: Discussion of different approaches and their regulatory use

October 9, 2007

08.30–09.30	Review of methodology and experiences with TME
09.30–12.00	Second breakout group session: Recommendations for performance and interpretation of TME studies 1
	Group 4: Fate and exposure considerations in design of TME
	Group 5: Ecological and effects considerations in design of TME
	Group 6: Statistical considerations in design of TME
12.00–13.00	Second plenary session: Recommendations for performance and interpretation of TME studies 1
14.30–17.00	Third breakout group session: Recommendations for performance and interpretation of TME studies 2
	Group 7: Sampling and determination of effects
	Group 8: Ecological interpretation and assessment of results
	Group 9: Potential regulatory use of TME endpoints
17.00–18.00	Third plenary session: Recommendations for performance and interpretation of TME studies 2

October 10, 2007

08.30–10.30	Reports of all workgroup rapporteurs
11.00–13.00	Synopsis and final plenary discussion

Appendix 2: Draft Method for Terrestrial Semi-Field Tests (Indoor and Outdoor Terrestrial Model Ecosystems)

This draft method reflects only an initial stage of recommendations on how to perform semi-field studies and needs to be developed further. It does not represent an endorsement by the PERAS workshop that the terrestrial model ecosystem (TME) approach is generally the preferred method for conducting semi-field studies. So far, the most experience in semi-field tests has been gained using TMEs that have been shown by several research groups to produce sound and reliable effect data of pesticides on soil organisms.

INTRODUCTION

Outdoor or indoor terrestrial model ecosystems are useful in risk assessment when lower-tier and higher-tier laboratory studies (single species or multispecies) indicate potential risks caused by plant protection products (PPPs). They can be an important tool in bridging the gap between these lower-tier studies and attempts to understand, predict, and confirm what may occur under field conditions. Every terrestrial model ecosystem study should be designed to test a specific hypothesis using information gained in previous tiers of the risk assessment. This makes every terrestrial model ecosystem study unique in at least some aspects of its design. Guidance for conducting terrestrial model ecosystem studies is therefore necessarily generic and flexible.

Although this draft method is mainly based on experiences with testing of PPPs, in principle, it could apply to other groups of potentially toxic substances (e.g., industrial chemicals). However, it should be noted that the exposure design for substances other than PPPs can differ (e.g., in relation to contaminated sites).

One important reason to perform a terrestrial model ecosystem study is to obtain more knowledge about the ecological relevance of effects identified in laboratory studies. The studies can therefore include a variety of species or functional endpoints. Interpretation of these studies focuses on effects at the population, community, and ecosystem levels, including indirect effects and the potential for recovery of affected organisms. A second important reason for conducting a terrestrial model ecosystem study is to measure effects of the chemical under more environmentally realistic exposure conditions.

The scope of this draft method mainly relates to determining environmental effects of PPPs. Fate properties are not specifically addressed in this draft method. However, the exposure of organisms should be confirmed by chemical analysis (preferably directly in the soil, but as a minimum by measuring the concentration of the PPP in the applied medium) at appropriate time intervals. Exposure can be determined by measuring either the total content in the soil or the concentration in the soil pore water. Again, as much relevant information as possible should be obtained and used from lower-tier environmental fate and behavior studies.

Concerning organism groups and endpoints to be studied for an effect assessment, soil mesofaunal groups such as nematodes, enchytraeids, collembolans, and mites as well as microorganisms are recommended. In detail, community and trophical structures as well as the abundance of these organisms are considered to be the most suitable endpoints. Optionally, the macrofauna (in particular earthworms) and functional endpoints (e.g., organic matter decomposition, feeding rates, microbial processes) can be included. Besides focusing on effects, it is possible to measure the intrinsic recovery of the whole system.

This draft method aims to describe the performance of TME studies in the context of the registration of pesticides. While not the main aim of this document, it is also possible to perform such studies at sites that were chemically contaminated in the past with pesticides (including aged residues). In such a case, concentration-response relationships can be mimicked by taking soil cores along a transsect from high to low exposure.

INITIAL CONSIDERATIONS

In principle, the relevance of the test system for the protection goals in question has to be discussed when planning the experiment and in the context of data interpretation. Before any terrestrial model ecosystem test is conducted, clear objectives should be defined in order to determine the relevant endpoints and which experimental design (e.g., level of replication, number of treatments) is appropriate. It is the responsibility of the study director to demonstrate that the system is appropriate for achieving the objectives of the study. It may be useful to discuss and agree on the protocol with the relevant authorities evaluating the test results. Any available information should be carefully reviewed, and preliminary laboratory testing should be undertaken when essential information for test design is missing. Factors to be considered include:

1) Effects: The core ecotoxicological data that are always required for registration and other higher-tier studies (e.g., additional single-species tests or population level studies) can be used to define the primary concerns to be investigated. For example, data on the sensitivity of soil species, knowledge on the pesticidal mode of action, and even existing lower-tier data on the effects toward terrestrial non-target arthropods can help to focus on those populations and communities that should be studied in more detail.

2) It should be defined what the derived parameters of interest are for each endpoint (e.g., ECx or NOEC).

3) The level of precision that is to be obtained for derived estimates, or the desired power of a relevant hypothesis, should be defined as part of determining the objective of the study.

4) The size of effects that are considered of ecological significance should be defined, relative to the endpoints of concern and the characteristics of the species, such as generation time and reproductive and migratory (recovery or recolonization) endpoints.

5) Depending on the objectives following from these 4 points, it will be possible to define:

   a) number of treatments and choice of doses,

   b) how treatments will randomly be assigned to the terrestrial model ecosystems,

   c) number of replicated terrestrial model ecosystems per treatment,

   d) organisms to be sampled, the size of the sample, and how sampling should be carried out, and

   e) the overall likely study duration and the numbers and timing of samples required (a reserve sampling option is recommended in case of a need for additional sampling during the study, or extension of the study duration).

6) The method of statistical analysis should be defined as part of the setting of the objectives (Chapman and Maund 1996), and a statistical design should be developed that ensures a desired statistical power.

7) An appropriate exposure regime should be established in order to meet the objectives of the study. Questions that should be addressed include:

   a) What are the expected routes of entry of the PPP into soil systems (e.g., spray application or seed treatment)?

   b) What is the frequency and timing of the application of the PPP?

   c) What is the expected fate and behavior of the substance within the test system, and how will this affect sampling and analysis?

   d) What worst-case predicted environmental concentrations (PECs) in soil is the study expected to replicate, and over what soil depths?

8) Physical-chemical and fate properties (for example, solubility, vapor pressure, octanol-water partition coefficient, adsorption coefficients, and biodegradability and persistence) should be previously ascertained alongside biological information in order to select sampling times and identify ecological components at greatest risk. A valid analytical method for soil, the water, and the stock solution should be available before performing the terrestrial model ecosystem test.

9) Information on the use patterns of the product should be given.

10) Similar exposure and effects information on the likely major metabolites should be considered when designing the study.

PRINCIPLES OF THE TEST

By definition, the type of terrestrial model ecosystems discussed here are parts of natural soil ecosystems. They are established by directly sampling intact soil cores in the field, e.g., from grassland or crop sites. In grassland, mesofauna density and diversity are generally higher and more stable than on arable land, which makes it easier to detect significant effects on the mesofauna. The test system thus contains a naturally developed soil community with appropriate organisms, such as microarthropods, enchytreids, nematodes, and microorganisms. Depending on the aim of the study, it may be appropriate to add certain organisms from external sources.

Soil risk assessment for PPPs is usually aimed at the protection of functions and communities. Thus, it is desirable for the TME test system to be broadly representative for communities of soil organisms. In practice, however, any test system can only simulate parts of 1 whole ecosystem, and extrapolations have to be made when transferring the results of a terrestrial model ecosystem study performed with a specific combination of soil and organisms to other such combinations from other sites. Unless multisite studies are conducted, this extrapolation has to be based on a scientifically reasoned case making use of relevant environmental and biotic information. It is hoped that further generic knowledge will be gained to support such extrapolations. Extrapolation to ecosystem components that have not been tested is not possible (e.g., recovery of univoltine species, i.e., those with just 1 generation per year).

The application of the test substance might be different for persistent and nonpersistent pesticides. The field situation should be mimicked as closely as possible. For persistent pesticides the long-term (accumulation) PEC plateau should be considered in the top layer of the soil, followed by the application of the yearly rate according to good agricultural practice (GAP), depending on the aim of the test (Van Der Linden et al. 2008b). To avoid conducting the test at too low a plateau concentration, it is recommended to check in advance, with an appropriate regulatory authority, the exposure input parameters and assumptions, such as soil depth and crop interception, that will be used to calculate worst-case PEC_{soil} values.

Terrestrial model ecosystem studies should preferably be designed in such a way that a concentration-response relationship may be identified over a range of ecotoxicologically relevant concentrations, encompassing those concentrations that reflect exposure in the field, e.g., PEC_{soil}. If the focus of the assessment is toward an expected endpoint (e.g., NOEC, ECx, or limit value), then the number, spread, and range of test concentrations, and the replication at particular concentrations, could be adjusted to improve accuracy and reliability. It should also be considered and accounted for in the choice of test concentrations whether any assessment/uncertainty factor is likely to be applied when using the endpoint in risk assessment.

Determining structural endpoints is the main aim of such a study. These structural endpoints relate to the abundance and biomass of all populations and their spatial, taxonomic, and trophic organization. Functional aspects could be documented as conditions of the study rather than as endpoints of the study, i.e., nutrient levels, respiration rate, mineral concentrations, pH, alkalinity, and organic material content. The terrestrial model ecosystem study should focus on taxonomic groups that, for example, in lower-tier risk assessments, have been identified as being of concern, as structurally or functionally important, or as exhibiting sensitivity to the test substance. In case the available database is limited, a corresponding thorough literature search should be performed.

Determining rate and extent of recovery of affected taxa can be crucial in the design of terrestrial model ecosystem studies. Looking at recovery is one of the key differences between terrestrial model ecosystem studies and other higher-tier studies and requires substantial ecological knowledge to interpret. If looking at recovery is an objective of the study, the experimental design should be such that recovery can be observed. A sufficiently long posttreatment period has to be foreseen to allow the detection of repopulation, e.g., an experimental duration of up to 1 year.

The data handling and statistical methods that are to be used to analyze the data should be built into the design of the study.

VALIDITY OF THE TEST

Because of the low level experience on the one hand and the complexity and variability of the test systems on the other hand, this draft method does not provide validity criteria. However, the validity of a study can be evaluated in light of the following conditions:

- Ideally, when concentration-response is studied, a clear effect level for at least the organisms of concern should be included, and at least 1 concentration that causes no effects that are considered ecologically significant (based on expert consideration of ecological function and recovery).
- Variability should be as small as needed to achieve the desired statistical power. If the variation between replicates is high, then the conclusions drawn from the study are less robust. By increasing the number of replicates the statistical power to address variability is increased.
- The amount of test material applied and the concentration in the spray solution have to be confirmed analytically at the start of exposure.

DESCRIPTION OF THE METHOD

TME FACILITY

The terrestrial model ecosystem facility can be constructed indoors as well as outdoors from any inert material, e.g., plastic or stainless steel. In order to prevent migration into the soil cores from outside the model ecosystem, the containers should be sealed at the bottom with a mesh of appropriate size. The single containers should be separated by an appropriate distance from each other, considering also practicability, e.g., size of the TME facility. In order to avoid migration of organisms into test containers, a minimum space of about 20 cm between containers should be given, and the space between containers should be filled, e.g., with gravel.

REUSE OF MODEL ECOSYSTEM CONTAINERS

Reuse of model ecosystem containers (without soil) after treatment with toxic chemicals depends on chemical characteristics, particularly the persistence of the chemical. For nonpersistent chemicals there may be no problem if it can be demonstrated that there are no longer any toxic residues present on the model ecosystem container. Alternatively, the containers can be drained and left empty for a sufficient period of time.

CORING SITE FOR THE TERRESTRIAL MODEL ECOSYSTEMS

Terrestrial model ecosystems can be sampled from either undisturbed grassland or arable land, depending on the intention of the study (even agroforestry sites are possible (Förster et al. 2006)). Since the density and diversity of the mesofauna living on grassland is higher than that on arable land, it should be easier to detect significant effects on soil cores from grassland. In general, the soil should be covered with plants. If application is only intended to bare arable soil, then plants should be grown subsequently, if feasible. The soil should be characterized by determining the particle size distribution (texture), water holding capacity, organic matter content, N and P content, cation exchange capacity, pH, and organic carbon content. The field history with respect to the chronology of treatment with PPP, fertilizer, and cultivation should be given for a reasonable timeframe, e.g., 3 years.

SOIL CORES

All soil cores used within a study should originate from the same site. In order to minimize the variation of the soil cores with respect to fauna and abiotic factors, coring should be conducted within a narrow area as much as possible. The coring process should be performed as gently as possible to minimize the impact on the consistency of the soil structure, and therefore also on the community of organisms living in the soil. The size of the cores may vary. Existing TME experience supports sizes of about 20 to 50 cm in diameter and a depth of about 40 to 60 cm. However, if required, different sizes are possible. Soil cores of larger diameters are considered to be more stable than smaller ones. The outcome of the study may consider intrinsic recovery from a possible impact of the test substance originating from the community present within the soil core (either natural field organisms or added organisms). Thus, soil cores of larger diameters may be more suitable than smaller ones.

Humidity of the soil is an important parameter for the community living in the soil. Therefore, this parameter should be recorded and has to be controlled. If the facility is outdoors, one can take advantage of the possibility of a natural control of the humidity regime via direct contact of the soil cores with the underground. However, in case of extreme conditions, e.g., drought, irrigation might be necessary. For indoor terrestrial model ecosystems the humidity has to be controlled either to simulate natural fluctuations or to create a "stable" environment.

ORGANISMS TO INCLUDE

The terrestrial model ecosystem study should focus on, but should not be limited to, taxonomic groups that lower-tier risk assessments have identified as being of concern. The test system is preferably a naturally developed soil community with appropriate organisms such as arthropods, nematodes, and/or oligochaetes. Besides known sensitivity, other selection criteria include ecological relevance, trophic level, and taxonomic position, in order to assess indirect or community level effects. The characterization of the actual community of organisms living in the soil should be done before the start of the study. Specifically, abundant species like oribatids and predatory mites, as well as collembolans, should be present in adequate number in the soil cores. To develop communities suitable to meet specific study objectives, it may also be acceptable to add organisms from appropriate external sources. TMEs may be considered for studying earthworm populations except for large vertical burrowers (anecics) if soil cores are large enough; however, it even may not be possible to sample more than one sample per soil core during a study.

Maturation Time

The period of time in which the terrestrial model ecosystems are adapted prior to chemical dosing should ensure the acclimatization of the community to the conditions of the testing facility. This period will take up to 2 to 4 weeks between coring and application of the test substance (see chapter research needs). To ensure homogeneous starting conditions, all soil cores should be characterized by presampling to exclude outliers 1 week before application.

Test Design

Appropriate test design depends on the purpose of the test: establishment of dose- (concentration-) response relationship, comparison with laboratory-derived NOEC, etc.

An exposure-response experimental design with replication allows a wider use of the data under different conditions and for different regulatory requirements. In this design, a terrestrial model ecosystem study should include at least 3 and preferably 5 concentrations, with at least 4 or 5 replicates per concentration, and 5 or 6 replicates as controls based on the available experience (Knacker et al. 2004). More concentrations may be required, depending on the slope of the dose-response relationship for the taxa of interest. The power to detect differences increases with more replicates, since replication reduces uncertainty in interpretation of results, and because test system variability can be better accounted for. The decision either to favor more replicates of each concentration (to calculate NOEC) or to prefer a concentration-response test design with less replicates and an increased number of concentration levels depends on the scientific questions to be answered. It is recommended to take at least duplicates from each soil core for the biological measurements.

In designing the terrestrial model ecosystem study, it may be helpful to consult a statistician to help determining which test design is required if an effect in a particular set of measured parameters is to be determined with a specific power. This will be a function of the replicate number and variability of the measurements. A design optimal for 1 variable will not necessarily be appropriate for another. The importance of focusing on critical endpoints cannot be overemphasized, e.g., the most sensitive group of organisms according to previous information.

Both uni- and multivariate statistics (e.g., principal response curve) should be employed for the evaluation of the study (e.g., Koolhaas et al. 2004; Moser et al.

2007). The NOECs should be reported for abundant single species as well as at the community level.

The selected concentrations should generally be based on those expected to cause effects. This should include the maximum predicted environmental concentration (PEC) (SANCO/10329/2002; EC 2002). Where relevant, multiple application regimes are possible. The selection of treatment levels should aim to include at least 1 concentration that will cause no ecologically significant effects and at least 1 that will cause clear effects. These concentrations can be derived from lower-tier studies or other sources, e.g., screening studies. The choice, particularly of maximum and minimum test concentrations, should be considered carefully against the expected PEC_{soil} values from current and future use patterns, and how the endpoint will actually be used in risk assessment, also considering the potential application of an assessment or uncertainty factor. If in doubt, it is recommended to check first with an appropriate regulatory authority.

Whether treatments should be assigned to experimental units at random or whether a constrained randomization should be employed, such as the arranging of treatments in replicate blocks, depends on the objective of the study.

Recovery

Determining the rate and extent of recovery of affected taxa can be an important factor in the design of terrestrial model ecosystem studies. When considering recovery, it is important to understand the potential influence of life history and dispersal mechanisms of the organisms involved, and possible interactions of these with the exposure regime and test system. Furthermore, in order to evaluate recovery, functional parameters (see Section 2.2) and possible adaptations and increased tolerance in the organisms and communities in response to the stress may be considered.

Another possibility to demonstrate recovery potential is to collect soil cores from the test system and to run a bioassay in the laboratory, e.g., to take soil samples at special intervals to add organisms in order to demonstrate the potential recovery of this species by colonization. However, it must be ensured that quality requirements usually used in laboratory tests are fulfilled.

PROCEDURES

Application

The test substance is added onto the test system after stabilization of the terrestrial model ecosystem. For pesticides, either the active ingredient (substance) or formulations may be used. The study and sampling design might be adapted to cover the potential effects of soil metabolites that appear within time, e.g., by application of persistent metabolites, if these are not expected to be formed at maximum levels during the natural course of the study, or by extending the study duration. For a more generic risk assessment, the use of the technical active ingredient may be preferred, unless it is difficult to work with the active ingredient, e.g., because of the lower solubility in water or if the formulation is much more toxic. If a formulation raises particular concerns due to, for example, its application as a solid rather than sprayed product (i.e., granule, bait, or seed treatment), then it might, in any case, be most appropriate to test the worst-case formulation. Regulatory objectives and properties of the active ingredient and formulations will help to determine which product should be used.

For persistent compounds, the baseline accumulated plateau concentration should be taken into account. Ideally the concentration should be established in the soil before additional application of the total annual application rate or dose, depending on the properties of the compound and the organisms of concern. However, by doing so through physical incorporation, the most important compartment for soil organisms would be destroyed. Alternatively, the plateau concentration may be included in the annual application rate, but this may cause an overdose of the test chemical, since the bioavailability of a freshly applied substance may be different compared to aged residues in the soil. For specific regulatory purposes, like the Dutch decision tree for persistence, other exposure regimes could be more appropriate, and a tailored design might be required.

Different approaches to applying the test substance to the terrestrial model ecosystem are available. However, the application should resemble the actual use identified in laboratory experiments (and according to the proposed good agricultural practice (GAP)) as being worst case. Possible exposure scenarios are, e.g., spray application, drench application, or application via treated seeds. In all cases, concentrations of the test substance established in the soil have to be analytically confirmed. The loading (amount of test substance added), timing, frequency of dosing, and number of replicates per treatment necessarily stem from the nature of the chemical, use patterns according to GAP, routes of entry, variability of the endpoint of concern, and objectives of the study. As in any experimental treatment using chemicals, care should be taken not to contaminate other terrestrial model ecosystems with test material.

Accounting for multiple application events causes a number of difficulties. It is advisable to apply an annual cumulative application in 1 dose on soil with only little plant cover or on bare soil. "Annual cumulative application" refers to the sum of all applications of the pesticide within a year. This should make no allowance for degradation of the test substance in soil, but accumulation in soil should be accounted for (see earlier). The crop interception levels for the applications at different growth stages should, however, be taken into account (see FOCUS 2000). In TME studies using grassland cores the interception of grass has to be taken into account and the grass should be closely mown or clipped prior to each application. However, clippings produced shortly before application should not remain on the test soil.

Multiple applications may also be performed in a study; however, this makes it more difficult to characterize the exposure and interpret the exposure-response relationships. For highly toxic but impersistent compounds, multiple applications according to the GAP may represent a higher risk of adverse effects with reduced opportunities for recovery.

The seasonal time of application of the test substance mainly depends on the degree of realism of the exposure scenario in the study. An application in spring or early summer usually is considered worst case, because most pesticides are used in this period. However, there may be a reason to apply a test system in autumn, for instance, if the pesticide is to be applied in autumn according to good agricultural practice. It is envisaged that a potential for recovery observed after an application in spring may have no relevance to the use of the product in autumn.

IRRIGATION

If no or little rainfall occurs within 3 days of the annual application, irrigation of the terrestrial model ecosystem is considered necessary to achieve optimal conditions for exposure. The amount used should be realistic according to regional and climatic conditions. A total of at least 10 mm (i.e., 10 L/m^2) of precipitation (rainfall plus irrigation) within 3 days of the spray application is desirable. This also applies for terrestrial indoor model ecosystem experiments in order to avoid a permanent surface spray on the vegetation or soil surface, but to guarantee a contamination of the soil layer. Additional irrigation, or the use of higher than usual spray water volumes, may help ensure that the test substance is washed down into the top soil layers rather than simply retained in the grass or leaf thatch and top root zone (this may be a particular issue for high log K_{oc} compounds).

SAMPLING

Useful information on ecological sampling for population and community level effects is available in the literature (e.g., Förster et al. 2004, 2006; Koolhaas et al. 2004; Moser et al. 2004a, 2004b, 2007; Römbke et al. 2004; Sousa et al. 2004; Kools 2006; Kools et al. 2009). As far as possible, standardized methods (e.g., ISO 2007a, 2007b, 2007c, 2007d) should be used.

It is recommended to assign measurements and samplings to specific locations in the terrestrial model ecosystem, for example, by the use of a positioning device. At each sampling, at least 2 replicates should be sampled in each TME in order to be able to extract different groups of soil organisms in different ways.

Pretreatment samples should be taken, for example, 2 weeks before application in order to assess and demonstrate the suitability of the test system. These pretreatment samples can also be used to perform covariate analyses in order to reduce residual variance among model ecosystems. More than one pretreatment sample may help determine natural fluctuations and population trends in order to put apparent effects and recovery into context. However, this will reduce the number of samples that can be taken during the exposure period. Sampling continues after treatment for the duration of the test. The total test duration and sampling regime depends on the aim of the study, the fate properties of the chemical, the life cycle, and recovery times of the populations of concern. Ideally, the study should continue long enough to demonstrate recovery of the affected species. Usually, a study duration of 1 year seems suitable in order to investigate long-term effects and recovery potentials over a full cropping season. However, recovery of some univoltine species could be missed unless the study is extended. A reserve sampling option is recommended in case of a need for an additional sampling during the study, or extension of the study duration.

The sampling regime during the exposure period also depends on the objectives of the test, the nature of the chemical, and the expected distribution of the chemical within the terrestrial model ecosystem. For example, in a test with a PPP known to cause immediate effects on certain organism groups in soil, the length of sampling intervals in the beginning of the test should be shorter than in the later phase. PPPs with a delayed effect on the community should be measured at longer intervals and for a longer period of time in order to demonstrate the actual effect of the PPP and the subsequent potential for recovery.

The sampling strategy has to ensure that the collection of samples does not change the terrestrial model ecosystem microenvironment. There are 2 sample strategies: either sampling the whole soil core and subsequent sub-sampling or taking subsamples from a soil core during incubation. If soil samples are taken from the model ecosystem, the remaining holes have to be refilled after sampling, e.g., with appropriate-sized plastic containers filled with, e.g., sand.

Biological Measurement

Microarthropods should be sampled by heat extraction using, e.g., a McFadyen extraction apparatus, a Berlese, or other suitable devices (e.g., ISO 2007b). Soil samples (of usually 5 cm diameter, 5 cm depth) should be taken from the terrestrial model ecosystem and as soon as possible placed in the extraction device. After applying an appropriate temperature gradient, e.g., starting from 25 °C up to 60 °C within 14 days, the microarthropods are collected and subsequently enumerated and identified using appropriate taxonomic keys, where possible to the species level. The temperature profile has to be evaluated before the start of the TME experiment, considering soil properties and the properties of the extraction device. If possible, the living animals should be identified down to the species level, employing appropriate identification keys.

Enchytraeids can be extracted from soil samples by, e.g., the wet extraction method. A possible procedure is described in ISO 23611-3 (ISO 2007c). Soil samples should be taken from the terrestrial model ecosystem, and as soon as possible the extraction should be started. If possible, the living animals should be identified down to the species level, employing appropriate identification keys.

Nematodes can be extracted from soil samples by various methods. A possible procedure is described in ISO 23611-4 (ISO 2007d). Soil samples should be taken from the terrestrial model ecosystem, and as soon as possible the extraction should be started. If possible, the animals should be identified down to the species level, employing appropriate identification keys.

Other organisms, like microorganisms, fungi, and some earthworm species (see earlier), can also be part of the terrestrial model ecosystem study. Ideally, standardized sampling procedures should be followed, e.g., ISO 23611-1 (ISO 2007a) for earthworm sampling.

Analysis of Test Chemical

The study objectives will determine the appropriate sampling and analysis strategy for the test chemical. There are 3 different reasons for doing a chemical analysis: to confirm that the test substance has been accurately applied to the test system, to quantify the chemical exposure initially and over time, and to relate it to the ecological responses observed.

The pesticide concentration in soil has to be measured by soil residue analysis. Subsamples of soil (usually 5 cm in diameter and 5 cm in depth) should be collected and analyzed immediately after incorporation of the plateau concentration into the soil by irrigation or rainfall. Three days after the application (if rainfall occurred) of the test dose, a set of soil subsamples should be collected. If irrigation is undertaken, soil samples for residue analysis should be collected 3 days after irrigation. Collection of soil samples for residue analyses should be performed according to standardized protocols, and standardized analytical methods to measure the pesticide should also be used where possible. In light of the wide analytical variability in field studies, it is recommended that a range of 50% to 150% of the nominal concentration should be reached (NAFTA 2005). At very low test concentrations, consideration needs to be given to the available limits of detection and quantification in soil.

DATA AND REPORTING

DATA HANDLING AND STATISTICAL ANALYSIS

Univariate analytical methods, such as analysis of variance (ANOVA) or regression analysis (or a combination of the 2, e.g., William's test), are best suited to investigate parametric data on effects at the population level of 1 species or taxon. The power of these methods to detect differences from the control response should be stated (Liber et al. 1992).

Multivariate analysis is appropriate for describing effects at the community level and can also be employed to indicate which taxa are particularly responsive to the treatment and would warrant specific univariate analysis. One method of multivariate analysis is based on the construction of principal response curves (Van den Brink and Ter Braak 1998, 1999) in which canonical coefficients describing the differences between the control and the treatments are plotted against time. This analysis takes into consideration the separate variances between replicates, between time points, and between treatments, thereby allowing clear representation of treatment effects in isolation. This pictorial evaluation of treatment effects can then be converted to a NOEC community with statistical significance using Monte Carlo permutation tests. Another method for evaluation of effects is the use of ecological models. Irrespective of the analytical tools applied for detecting the differences, the power of them should be considered. If the simulation type of approach was used, then the effect or the no-effect level should be related to loading.

Another method for evaluating terrestrial model ecosystem studies is by calculating diversity or similarity indices. Analogous to the multivariate method described earlier, nonparametric multivariate methods may also be used for the analysis of communities. Similarity indexes such as the Bray Curtis similarity index (Bray and Curtis 1957) may be coordinated for pictorial evaluation and statistically analyzed using Monte Carlo permutation tests effectively separating time and treatment effects (Clarke 1993; Clarke and Warwick 1994, 2001). Whichever method or combination of methods of analysis is to be applied, it is essential to build the analytical technique into the design of the study at the outset rather than to search for an appropriate method after the data have been generated.

It may turn out that some NOECs from individual species may be lower than the community NOEC. In that case, the ecological role and specific characteristics of that or those species and other related species should be considered. As mentioned previously, the magnitude, duration (i.e., recovery or reconciliation potential), and nature of particular effects on key species, populations, or other taxonomic, age class, or ecological groupings (or even functions) may need to be examined in more detail. Therefore, based on a careful scrutiny of lower-tier data, suitable analytical methods should be built in at the outset to achieve this.

APPROPRIATE LEVELS OF TAXONOMIC RESOLUTION

Those taxa that are most sensitive should be identified to the species level (where practicable and appropriate—some taxa may already give meaningful results at higher levels, like genera (enchytraeids) or trophic groups (nematodes)). Species level identification is recommended for other taxa where practicable, since it may permit effective use of multivariate statistical approaches for more powerful analysis of community structure. The level of taxonomic analysis therefore depends on the objectives of the study. Univariate statistics may not be sufficiently powerful to detect differences among groups of rare organisms. In this case, data may be aggregated appropriately into larger taxonomic groups before analysis (Giddings et al. 2002), provided this will not mask effects that could be seen using other approaches.

Other groups of organisms that are identified as less sensitive in lower-tier tests may be monitored less intensively (e.g., at a lower level of taxonomic resolution, such as family), although the ideal is species level identification. However, one reason to monitor these taxa is the possibility of the occurrence of indirect (ecological) effects (Frampton 2007). Identification of organisms should be possible with available taxonomic keys and without breeding larval or nymphal forms through to older stages.

REPORTING REQUIREMENTS

The final report should give a full and comprehensive description of the study, including its objectives, design, and detailed results for each species, function, or other grouping or index. Presentation of clear, tabular, pictorial, or graphical representations of the nature, magnitude, and duration of spatial or temporal effects can greatly assist with interpretation. Along with a description of the analytical and statistical techniques employed, the following data should also be reported, depending on the study approach and the objective of the study:

Information on test substance and relevant metabolites:
- Identification, including chemical name and Chemical Abstracts Service (CAS) number
- Batch or lot number
- Identification and levels of impurities
- Chemical stability under the conditions of the test

- Volatility
- Specific radioactivity and labeling positions if appropriate
- Formulation details, where applied as a formulated product
- Method for analysis of test substance and transformation products, including limits of analytical detection and quantification
- Physicochemical properties of the test substance, partition coefficients, rates of hydrolysis, photolysis, etc.

Test systems:
- Description of coring site, location, history, microarthropod community, and its special variability
- Description of test systems, location, history, dimensions, construction materials, etc.
- Soil quality: Description of the chemical and physical parameters of the soil used in the test system
- Description of variation between replicates
- Experimental design and measured data
- Treatment regime: Dosing regime, duration, frequency, loading rates, preparation of application solutions, application of test substance, etc.
- Sampling and analysis, residue monitoring results, analytical method
- Meteorological records at, or very close to, the test site
- Physicochemical soil measurements (humidity, etc.)
- Sampling methods and taxonomic identification methods used
- Biological results per species or taxonomic group
- Statistical methods used
- Outcome of the statistical evaluation

Appendix 3: Workshop Sponsors

We gratefully acknowledge the generous support of BASF, Bayer CropScience, the Dutch Ministry of Housing, Spatial Planning and the Environment, the IMAR–Coimbra Interdisciplinary Centre, Syngenta, and the German Federal Environment Agency (Umweltundesamt, UBA).

The Chemical Company

**IMAR-Coimbra
Interdisciplinary Centre**

**Dutch Ministry of Housing, Spatial
Planning and the Environment**

Index

99

Other Titles from the Society of Environmental Toxicology and Chemistry (SETAC)

Linking Aquatic Exposure and Effects: Risk Assessment of Pesticides
Brock, Alix, Brown, Capri, Gottesbüren, Heimbach, Lythgo, Schulz, Streloke, editors
2009

Derivation and Use of Environmental Quality and Human Health Standards for Chemical Substances in Water and Soil
Crane, Matthiessen, Maycock, Merrington, Whitehouse, editors
2009

Aquatic Macrophyte Risk Assessment for Pesticides
Maltby, Arnold, Arts, Davies, Heimbach, Pickl, Poulsen, editors
2009

Copper: Environmental Fate, Effects, Transport and Models: Papers from Environmental Toxicology and Chemistry, 1982 to 2008 and Integrated Environmental Assessment and Management, 2005 to 2008
Gorsuch, Arnold, Santore, Smith, Reiley, editors
2009

Veterinary Medicines in the Environment
Crane, Boxall, Barrett
2008

Relevance of Ambient Water Quality Criteria for Ephemeral and Effluent-dependent Watercourses of the Arid Western United States
Gensemer, Meyerhof, Ramage, Curley
2008

Extrapolation Practice for Ecotoxicological Effect Characterization of Chemicals
Solomon, Brock, De Zwart, Dyer, Posthuma, Richards, Sanderson, Sibley, van den Brink, editors
2008

Environmental Life Cycle Costing
Hunkeler, Lichtenvort, Rebitzer, editors
2008

Valuation of Ecological Resources: Integration of Ecology and Socioeconomics in Environmental Decision Making
Stahl, Kapustka, Munns, Bruins, editors
2007

Genomics in Regulatory Ecotoxicology: Applications and Challenges
Ankley, Miracle, Perkins, Daston, editors
2007

Population-Level Ecological Risk Assessment
Barnthouse, Munns, Sorensen, editors
2007

Effects of Water Chemistry on Bioavailability and Toxicity of Waterborne Cadmium, Copper, Nickel, Lead, and Zinc on Freshwater Organisms
Meyer, Clearwater, Doser, Rogaczewski, Hansen
2007

Ecosystem Responses to Mercury Contamination: Indicators of Change
Harris, Krabbenhoft, Mason, Murray, Reash, Saltman, editors
2007

Genomic Approaches for Cross-Species Extrapolation in Toxicology
Benson and Di Giulio, editors
2007

New Improvements in the Aquatic Ecological Risk Assessment of Fungicidal Pesticides and Biocides
Van den Brink, Maltby, Wendt-Rasch, Heimbach, Peeters, editors
2007

Freshwater Bivalve Ecotoxicology
Farris and Van Hassel, editors
2006

Estrogens and Xenoestrogens in the Aquatic Environment: An Integrated Approach for Field Monitoring and Effect Assessment
Vethaak, Schrap, de Voogt, editors
2006

Assessing the Hazard of Metals and Inorganic Metal Substances in Aquatic and Terrestrial Systems
Adams and Chapman, editors
2006

Perchlorate Ecotoxicology
Kendall and Smith, editors
2006

SETAC

A Professional Society for Environmental Scientists and Engineers and Related Disciplines Concerned with Environmental Quality

The Society of Environmental Toxicology and Chemistry (SETAC), with offices currently in North America and Europe, is a nonprofit, professional society established to provide a forum for individuals and institutions engaged in the study of environmental problems, management and regulation of natural resources, education, research and development, and manufacturing and distribution.

Specific goals of the society are

- Promote research, education, and training in the environmental sciences.
- Promote the systematic application of all relevant scientific disciplines to the evaluation of chemical hazards.
- Participate in the scientific interpretation of issues concerned with hazard assessment and risk analysis.
- Support the development of ecologically acceptable practices and principles.
- Provide a forum (meetings and publications) for communication among professionals in government, business, academia, and other segments of society involved in the use, protection, and management of our environment.

These goals are pursued through the conduct of numerous activities, which include:

- Hold annual meetings with study and workshop sessions, platform and poster papers, and achievement and merit awards.
- Sponsor a monthly scientific journal, a newsletter, and special technical publications.
- Provide funds for education and training through the SETAC Scholarship/Fellowship Program.
- Organize and sponsor chapters to provide a forum for the presentation of scientific data and for the interchange and study of information about local concerns.
- Provide advice and counsel to technical and nontechnical persons through a number of standing and ad hoc committees.

SETAC membership currently is composed of more than 5000 individuals from government, academia, business, and public-interest groups with technical backgrounds in chemistry, toxicology, biology, ecology, atmospheric sciences, health sciences, earth sciences, and engineering.

If you have training in these or related disciplines and are engaged in the study, use, or management of environmental resources, SETAC can fulfill your professional affiliation needs.

All members receive a newsletter highlighting environmental topics and SETAC activities and reduced fees for the Annual Meeting and SETAC special publications.

All members except Students and Senior Active Members receive monthly issues of Environmental Toxicology and Chemistry (ET&C) and Integrated Environmental Assessment and Management (IEAM), peer-reviewed journals of the Society. Student and Senior Active Members may subscribe to the journal. Members may hold office and, with the Emeritus Members, constitute the voting membership.

If you desire further information, contact the appropriate SETAC Office.

1010 North 12th Avenue
Pensacola, Florida 32501-3367 USA
T 850 469 1500   F 850 469 9778
E setac@setac.org

Avenue de la Toison d'Or 67
B-1060 Brussels, Belgium
T 32 2 772 72 81   F 32 2 770 53 86
E setac@setaceu.org

www.setac.org
Environmental Quality Through Science®

Printed and bound by CPI Group (UK) Ltd, Croydon, CR0 4YY

21/10/2024

01777084-0020